ROASTING

OF

GOLD AND SILVER ORES,

AND THE

Extraction of their Respective Metals

WITHOUT QUICKSILVER.

By G. KUSTEL,

MINING ENGINEER AND METALLURGIST,

Author of " Nevada and California Processes of Silver and Gold Extraction," and
" Concentration of all Kinds of Ores."

NEW EDITION.

ILLUSTRATED WITH NUMEROUS ENGRAVINGS.

A. J. Leary, Stationer and Printer, 402 & 404 Sansome St.
San Francisco, 1880.

PREFACE TO THE SECOND EDITION.

The favor with which the first edition of the present treatise on Roasting of Ores was received by the mining public, and the increased inquiries after a new edition, induced the author to write the present book.   During the long interval of ten years which elapsed since the first edition, many improvements in the construction of furnaces, as well as in the way of roasting, have been introduced; this edition, therefore, has undergone many alterations, and new cuts of furnaces, etc., have been added.

San Francisco, Feb., 1880.   G. K.

PREFACE TO THE FIRST EDITION.

The publication of this Treatise is due solely to the many inquiries concerning the "Leaching, Solving and Precipitation Process for Silver Ores," now successfully practiced in Sonora, Mexico, where it has been lately introduced by Mr. Ottocar Hofmann.

In consideration of the very important preparation of the ore, before it is subjected to the Solving Process,—namely, the Roasting,— I have thought it proper to devote considerable space to the description of different modifications of this operation, which is regulated by the peculiarity of the ore, and by the subsequent treatment. It is impossible to give any one way which will be suitable in every case; for this reason, and in order to cover all cases as far as possible, a detailed description of different modes of Roasting will not be superfluous.

The Solving Process, as now practiced, is a very economical method for the extraction of silver, for the reason that no quicksilver and no castings are used except what are needed for crushing. Mills in Mexico being dependent on San Francisco for the shoes, dies, gearing, etc., of amalgamating pans, millmen there know how to appreciate a process confined to wooden tubs requiring no power. A comparatively small capital is necessary for building up such works, and hence there is a more reasonable ratio established between the amount of money which must be expended on the works and the real value of the mine, than where other more expensive machinery is employed—a circumstance which, being insufficiently regarded, is often the source of failure.

Mr. O. Hofmann commenced first with the "Chlorination Process" but finding great difficulty in obtaining the regular supply of sulphuric acid and manganese from San Francisco, abandoned the chlorination with cold chlorine gas, which is indispensable in the presence of gold. Another difficulty was in obtaining a good article of sulphide of sodium. He tried to extract the potash from ashes, and to use this in place of soda, but decided finally in favor of lime, which is found in abundance. From this the sulphide of calcium is easily manufactured on the spot. Sulphide of calcium was first applied by Kiss.

The Solving Process is very simple, and readily performed by common workmen; besides the lime, only brimstone must be provided, in order to prepare the necessary chemicals for solving and precipitation. It is a general, but erroneous, belief, that the solving is a slow process. An amalgamating pan is charged with 500 to 1,000 pounds of roasted ore, and treated at least six hours, and therefore turns out at most two tons in 24 hours; while a box or vat of proper size used in the Solving Process, can work from four to ten tons in the same time.

Only those ores are treated by this process which absolutely require roasting; which, however, with improved furnaces, is not so expensive as it used to be. The chloride ores alone can be leached directly without roasting, and this when there is no other silver combination in them.

G. Kustel.

March, 1870.

INDEX.

I.

III.

IV.

V.

I. INTRODUCTION.

Ores are classified: *a.* According to the metal, the extraction of which is principally remunerative; as silver ores, lead ores, copper ores, etc. *b.* According to the metallurgical treatment; as roasting ores, smelting ores, milling ores, etc. *c.* According to the predominant gangue; as calcareous ores, quartzose or ochery ores. *d.* According to the predominant metallic mineral; as sulphuret ores, chloride ores, carbonate ores, etc.

Important Silver Ores.

The most important silver ores are those found in such quantities as to be an object of metallurgical operations. The principal minerals of this kind are the following:

A. Silver Ores with unvariable amount of Silver. *a. Sulphuret of Silver,* or silver glance, with 87 per cent. of silver. It is of common occurrence, and is the most suitable of the silver sulphurets for pan amalgamation without roasting. *b. Brittle Silver Ore, (Stephanite)* or sulphuret of silver and antimony. This mineral contains 68 per cent. of silver, and is quite common. *c. Polybasite,* sulphuret of silver, antimony and some arsenic, with 75 per cent. of silver. Horn Silver, or chloride of silver, with 75 per cent. of silver, occurred massive in White Pine, Nevada, Arizona; iodic and bromic silver of yellow and green color. *d. Ruby Silver.* The dark red silver ore, or antimonial variety, with 59 per cent., and the light red silver ore, or arsenical variety, with 65 per cent. of silver, are valuable minerals. They occur quite frequently in Nevada, Idaho, Montana,

Mexico, etc.   *e. Miargyrite*, sulphuret of silver and antimony; 36.5 per cent. of silver; Idaho, Montana.

B. Argentiferous Ores, with variable amount of silver; as *Stromeyerite*, or silver copper glance, a sulphuret of silver and copper containing up to 53 per cent. of silver; beautiful specimens are found in the Silver King and Heinzeman mines, Arizona. *Stetefeldtite*, with 25 per cent. of silver, is an oxide ore which is found very frequently in Nevada and Arizona.   *c. Silverfahlore*, argentiferous gray copper ore.   It contains silver in very variable proportions up to 31 per cent.   This ore is quite common, and for this reason is important.   It is also one of the most rebellious ores, containing copper, antimony, arsenic, sulphur, lead, iron, zinc, and sometimes gold and quicksilver.   *d. Chloride Ores*, (so called) mostly decomposed ores, generally of an earthy appearance and different color.   They contain more or less finely divided chloride of silver.

C. Argentiferous Lead Ores, galena, or sulphuret of lead, lead glance.   Generally, this is not rich in silver, containing from $20 to $60 per ton.   Specimens assay sometimes as high as $300.   The fine-grained variety is generally considered richer than the coarse crystallized kind, but this has not been observed to be the case in Nevada and Arizona.   *c. Cerusite*, carbonate of lead.   If pure, without admixture of copper and other carbonates, it is poor in silver in most cases.   Raw, it amalgamates only too readily in pans.   Smelting is the only proper way of treating galena and cerusite.   *d. Argentiferous Zincblende*, sulphuret of zinc.   Pure zincblende contains usually only traces of silver; often, however, it assays well, even up to $400 per ton, although no other silver ore can be detected with it. In some mines the argentiferous zincblende prevails, and is the most important ore.   It requires a great heat in roasting.   *e. Argentiferous Pyrites*.   Copper and iron pyrites are poor in silver, but often auriferous.   Pyrite is a valuable companion for silver ores which have to be treated by a chloridizing roasting, on account of its amount of sulphur, which is necessary for the decomposition of salt.

There are, besides, numerous classes of decomposed silver ores, generally of earthy nature ; also, half decomposed ores which have lost their metallic glance, having a black or bluish-black color, and being generally cupriferous.

Free Milling Ores.

Free milling means the amalgamation of ores in pans directly, without roasting. The decomposition of the ore is effected by the addition of chemicals (blue vitriol and salts). Of all the minerals above stated, only the silver glance, chloride ores, bromic and iodic ores, are free milling ores. Less favorable results by free milling are obtained from stetefeldtite, from some of the decomposed ores, from silver—copper glance (if rich in silver), and from stephanite. All other sulphureted and many decomposed ores, especially those rich in antimony or arsenic, cannot be treated in pans, or by lixiviation, without a proper roasting.

Difference between Real Silver Ores and Argentiferous Ores.

Real silver ores have mostly an unvariable amount of silver. Real silver minerals admit an approximate estimate of the value of the ore, if the proportion of ore and gangue is considered, without making an assay. For instance, if ruby silver ore should consist, according to guess, of one part of ruby and about three parts of quartz. or some other gangue, then one ton would contain $\frac{1}{4}$, that is, 500 lbs. of ruby, which always assays 59 per cent. of silver ; therefore, one ton of such ore would be estimated as containing 4,301 ounces (troy) of silver. With the argentiferous ores it is different. Fahl ore, for instance, may be very poor or very rich, yet its value can be ascertained only by an assay. There are no means of estimating the richness of argentiferous ores "by sight."

Important Combination.

With the exception of a few metal oxides of iron, zinc, manganese, and, among silver ores, of the stetefeldtite, etc., the most important, because the most frequent ores, are the sulphureted varieties. Sulphur is the most formidable obstacle to the metallurgist in extracting metals from their respective ores. Desulphurization has been a subject of most diligent and numerous experiments. The oldest method is the application of heat, which is still in use, notwithstanding the many attempts in modern times to dispense entirely with fire or to modify its application so as to perform the process more perfectly and in a shorter time. The only important progress in desulpurization by heat in a very short time, has been made with the Stetefeldt roasting furnace.

Means of Desulphurization.

The desulphurization of ores is effected : a. By heating with free admission of air. This is the common way of "roasting," and the most important, and is effected either in kilns, heaps, etc., or in reverberatory furnaces. As soon as the sulphureted ore is heated to a certain degree, one part of the sulphur escapes as sulphurous acid ; another is converted into sulphuric acid. Some sulphurets (iron pyrites) lose their sulpher without the application of heat, being decomposed by exposure to the action of air for a long time. This way is sometimes practiced on gold-bearing pyrites. b. By heating with exclusion of air. Only the sulphuides of gold and platinum are decomposed perfectly by this method. Other sulphureted ores lose their sulphur only in part, being reduced to a lower state of sulphide.

Sulphuret of silver (Ag S) remains undecomposed. Cinnabar, sulphide of antimony (Sb S^3) and sulphide of arsenic volatize unchanged. Iron pyrites (Fe S^2) gives up 23 per cent. of its sulphur, being reduced to magnetic pyrites, and, by a strong heat, to protosulphide of iron (Fe S), not further reducible. Also sulphide of zinc (zincblende), remains undecomposed. Copper glance retains its sulphur, and copper pyrites loses only one part of the sulphur which is combined with the iron in it. Galena (Pb S) is reduced to a lower state (Pb^4 S), a part of the lead separating out in a metallic state. (Fb^4) c. By superheated steam. Sulphurets not evolving sulphur by the last process, lose their sulphur slowly on the application of steam, sulphureted hydrogen and sulphurous acid being formed. Experiments made by Regnault showed that desulphurization is effected more perfectly if air is admitted. Roasting in reverberatory furnaces is always effected by the oxygen of the air and by steam, as there is no fuel used which contains less than 25 to 30 per cent. of water. Superheated steam has been tried in different ways on sulphurets with the highest expectations, but with no better results for practical use than are given in the ordinary way by the steam obtained from fuel. It may be useful in many instances to have more steam than is thus obtained, but this increases considerably the expense of roasting; as, for instance, in Patera's application of steam in roasting silver ores, tried principally with the intention of expelling antimony, arsenic, etc. Another application of superheated steam, with exclusion of air, is Hagan's method, which may prove sucsessful on pyritous ores, having at the same time the advantage of being a very cheap method. d. By heating with metals, alkalies or alkaline earths, for which the sulphur has a greater affinity. The affinity of sulphur for the following metals decreases in the order in which they stand, being strongest for the first and weakest for the last: Copper, iron, tin, zinc, lead, silver, antimony, arsenic. Each of these metals can be desulphurized by the next preceding, though with difficulty;

but more easily by one further off. Practical use of this property is made in smelting galena with the addition of metallic iron or iron ore. Sulphide of silver in crucibles is decomposed by stirring the liquid with red hot iron. In a cold way silver sulphurets and chlorides are decomposed by iron in amalgamating pans, the chlorides by iron and quicksilver. Quicksilver is obtained from cinnabar by heating the latter with lime, which takes up the sulphur, etc. *e.* Carbon has no great affinity for sulphur; the use of charcoal for desulphurization of ores is therefore an inferior method. So is also the use of carbonic acid.

Result of Desulphurization.

The direct extraction of metals from sulphurets, either by smelting or amalgamation, is not practicable. In smelting, the sulphurets melt very readily, but only a small part of the metal is obtained, while the greater part runs out combined with the sulphur as matte. For this reason the roasting of sulphureted ore for the purpose of smelting is indispensable unless iron is added, which desulphurizes the ore. Such roasting or burning takes often many weeks or months. The direct amalgamation of sulphurets gives a very poor result, except in the case of silver glance. By means of the chemical action of sulphate of copper and salt, the silver and gold sulphates are decomposed; but no process has yet been publicly demonstrated as really practical for the treatment of all kinds of raw sulphurets. The desulphurization is therefore still a most important preparation for the extraction of metals. The general effect of roasting is that the metals become oxidized. Only gold and silver are transformed into a metallic condition; and of the silver, moreover, a large percentage is always found as a sulphate, even when the roasting is well performed. Some of the silver combines as an oxide with antimony and silica, if present. All the oxides obtained by desulphurization must be again deoxidized in order to get them in a metallic state.

Means of Reduction or Deoxidation.

Heating alone will reduce the oxides of the precious metals only. Oxide of gold does not occur in nature, neither is it obtained in any of the metallurgical processes. Oxide of silver is also unimportant; it is formed, to a small extent, in cupellation (and taken up by the litharge), in smelting silver ores combined with silica, and in roasting silver ores in the presence of antimony, arsenic, etc.

The most powerful agent of reduction is carbon (charcoal, coke, etc.,) and carbonic oxide. In all smelting in blast furnaces, the carbonic oxide is the real reducer. The burning coal, under the influence of the compressed air, produces carbonic acid, melting at the same time the ore; the carbonic acid, passing through the glowing coal above the melting region, gives up a part of its oxygen to the coal, and is reduced thereby to carbonic oxide, which in turn takes up oxygen again, from the metal oxides, reducing them to a metallic state; a contact of ore oxides with carbon is therefore not necessary for the purpose of reduction. All metals do not retain their oxygen with equal tenacity, but some part with it much more easily than others. For instance, lead, copper, bismuth, antimony, cobalt and nickel, require for their reduction a darker or lighter red heat; while iron, zinc and tin are reduced only at a white heat. But also hydrogen and carbureted hydrogen, created by the burning fuel, are powerful reducing agents.

Metal oxides in solution are reduced and precipitated in a metallic condition by other metals. On this principle copper is precipitated by metallic iron, which goes into solution in place of the copper; sulphate of silver, in Ziervogel's process, is precipitated by copper, etc. Also, by aid of electro-galvanic stream, metals are reduced to a metallic state from their solutions.

Desulphurization of Silver Ores not Efficient.

Although by mere desulphurization the silver is to a great extent converted into a metallic state, this is not always its most suitable condition except for smelting. The larger part of all silver extracted in the United States is obtained by amalgamation, smelting being confined to localities where the ore contains such a high percentage of lead that its amalgamation is impossible. It would seem as if metallic silver should amalgamate more easily than if combined with another substance. This, however, is not the case. The silver, after roasting, is generally coated with the oxides of volatile base metals, which prevent its ready amalgamation. Moreover, a direct contact between quicksilver and silver is a necessary condition for their amalgamation. A momentary contact in a muddy pulp is not always successful. The chloride of silver, however, goes into solution and unites easily with the quicksilver. Hence, in most instances, it is necessary to adopt a chloridizing roasting.

What a Chloride is, and how Chlorination is Effected.

The term *chloride* is applied to all compounds of chlorine with a metal or other radical. Chlorine is a greenish-yellow gas, an elementary substance, of 2.45 specific gravity, and of a peculiar and disagreeable odor. It is not found free in nature, but always in combination, principally with sodium, forming common salt. Metallic chlorides are of frequent occurrence. Chlorine is, for instance, combined with silver as horn silver, with copper as Atacamite, with lead as Kerasine, Mendipite, etc.; also with quicksilver as Calomel.

To chloridize ore, that is, to convert the metals into chlorides, it is necessary to produce chlorine, and to bring it in intimate contact with the ore particles. The cheapest material evolving—chlorine is salt (chloride of sodium), and the only practical way of separating the chlorine from sodium is by substituting for it another substance for which the sodium has a stronger affinity. The cheapest ingredient for this purpose is sulphuric acid. The sodium being oxidized to soda, unites with the sulphuric acid, forming sulphate of soda, while chlorine is set free.

For the treatment of ores there are two principal methods of chloridizing. One is roasting the ore with salt in a furnace; the other is the "cold chlorination." Roasting, at first, when in the presence of salt, has an oxidizing effect, as there is then no sulphuric acid present to decompose the salt, and the heat alone would, if increased, volatilize, and only imperfectly form sulphuric acid. The sulphurets in the ore, under the influence of heat, lose a part of their sulphur as sulphurous acid gas; the other part of the sulphur oxidizes to sulphuric acid. As soon as this is formed it attacks the salt, and the chlorine, being set free, then acts on metals, metal oxides, sulphurets, arseniurets and antimonial combinations, forming partly metal chlorides and partly chlorides of sulphur, arsenic and antimony.

The other mode of chloridizing consists in the employment of cold chlorine gas with roasted ores, principally desulphurized gold ores. The chlorine must be produced here separately, and conducted into the cold ore by leaden or india rubber pipes. The ingredients are: Salt, sulphuric acid and peroxide of manganese. Salt is first attacked by the sulphuric acid, and hydrochloric acid and sulphate of soda are formed. The hydrogen of the hydrochloric acid then combines with the oxygen of the manganese, and the chlorine escapes. A part of the chlorine unites with the manganese, but is decomposed again by sulphuric acid, so that all chlorine is expelled from the salt, leaving sulphates of soda and manganese in the gas generator. The chlorination of gold, unlike that of silver, is difficult to effect in a furnace

(Pg42) for the reason that, if formed, the gold chloride is reduced back
to the metallic state at a low, almost dark red heat. The difference between hot and cold chlorination is principally found in the fact that,
while in the first way a great many base metal chlorides are formed,
the cold chlorine combines principally with the free metal, with silver
and gold; while the other metals, being oxidized, are not decomposed
by the chlorine. Silver oxide, if present, is decomposed and chloridized.

Chlorination is also effected by chemical decomposition in the wet
way, as practiced in the Mexican patio amalgamation, by mixing
with the ore sulphate of copper and salt.

Means of Separating the Metal from Chlorine.

The chloride of silver can be melted without being altered; chlorides of gold and of platinum lose all their chlorine on being heated.
In this way, the gold, not uniting with the chlorine in the heat, can
be perfectly refined and separated from other metals in a very simple way, by introducing chlorine gas through a porcelain or earthen
pipe into the liquid gold, while in the crucible. The silver, copper
and other base metals unite with the chlorine and rise to the surface
as chlorides. Chloride of iron exposed to air and heat, as is the case
in a chloridizing roasting, loses its chlorine and is changed to iron
oxide. The chloride of copper gives up only a part of its chlorine.
Heating alone has therefore no practical value for the disengagement
of chlorine.

The most effective way of separating the chlorine from the metal
is the application of another metal for which the chlorine has more
affinity. On this property of chlorine is based the amalgamation of
silver ores after a chloridizing roasting in pans, tubs and barrels, and
the patio amalgamation. The chloride of silver in the ore is decomposed, and the silver set free during amalgamation in iron pans by

the metallic iron of the pan, or if quicksilver is charged at the same time with the ore, by both the quicksilver and the iron. In the barrel amalgamation the silver is disengaged by metallic iron, and in the patio amalgamation by quicksilver. In all these instances the silver, being deprived of its chlorine, alloys with the quicksilver and forms the amalgam.

On the same principal the metal is extracted from soluble chlorides. The proto-chlorides are all more or less soluble in water, except that of silver, which is quite insoluble. The chloride of copper, in solution, is brought together with metallic iron, or conveyed over it. The chlorine of the copper unites with the iron, and the copper falls in a metallic state, ready to be melted into bars, after being washed, pressed and dried. Indirectly, the silver is extracted from its chloridized state by dissolving the chloride in the hyposulphites of soda, of potash, or of lime. In these salts the silver chloride dissolves very readily, giving a clear solution of a very sweet taste, out of which the silver is precipitated by the corresponding alkaline sulphides as sulphide of silver.

The chloride of gold, obtained from the chlorination of gold-bearing sulphurets, is precipitated by sulphate of iron in such a way that metallic gold results, while the chlorine combines with a part of the iron.

The silver is easily obtained from the chloride by melting it with alkalies; for instance, with soda, potash or lime. The chlorine unites with sodium, calcium, etc., and the silver separates on the bottom of the crucible. If there is not a sufficient amount of the alkalies present, some silver will be lost. In most instances it is preferable to mix the artificial chloride with water and some sulphuric acid and granulated zinc, or zinc sheet if smaller quantities are being operated on. The chloride of silver by degrees changes its white color to a dark gray, being converted into the metallic state in a short time. It is reduced to metal by the nascent hydrogen. After the sulphate of zinc, which is formed and dissolved, has been washed away, the silver is pressed, dried, and, with addition of some soda and borax, melted into a bar.

In the same way as from a sulphate, silver can be precipitated by copper, after the chloride of silver has been dissolved in a hot solution of salt, as is done in Augustin's process. This is not practicable with the argentiferous solution of hyposulphite of soda.

By using sodium amalgam and iron filings, the silver chloride is instantly decomposed and silver amalgam formed.

The chloride of gold is precipitated in a metallic condition by the chloride of iron (Fe Cl), the consideration of which is important in treating sulphurets by chlorination.

II. ROASTING OF ORES.

The object of roasting is either to effect chemical changes, as required for amalgamation, smelting, etc., or to reduce the hardness of the ore, in order to make it easier to crush. Roasting for the latter purpose, exposing the ore to the fire in large pieces, is more properly termed "burning." The beginning of smelting is under all circumstances beyond the limits of roasting; therefore all roasting furnaces in which the regulation of heat is so far out of the control of the roaster that a partial smelting would arise, are unfit for roasting. This is often the case with vertical furnaces. But although a partial smelting or clotting is not within the province of roasting, and in all instances is very injurious to the result of subsequent amalgamation or precipitation, it is nevertheless applied with much success on concentrated ore intended for smelting. By this process the loose sand assumes a compact form, the gases and wind penetrate the charge more easily, and the loss in metal is diminished.

If there is no necessity for effecting a perfect chemical change in the ore, or if roasting is required for smelting purposes, and a powdered form is not admissable, the ore is taken in larger or smaller pieces—generally not below the size of a hen's egg—and subjected to roasting either in open heaps, in kilns, or in vertical or reverberatory furnaces. In roasting in heaps, the wood is first placed on the ground, sometimes surrounded by a wall two or three feet high, then the ore is put over it. Less frequently ore and wood are laid in strata. If there is sufficient sulphur in the ore, the burning will continue without addition of fuel for many days or weeks. It is evident that the result of such roasting is very unequal, the outside being more oxidized than the idside, the heat greater near the fuel than further off, etc. For this reason such ore is often roasted over several times.

In vertical furnaces the ore is laid in strata alternating with fuel, or there are several fire-places outside the furnace so arranged that the flame is conducted by the draft into the furnace. A modification in construction and principle is the Hagan roasting furnace, in which the decomposition of superheated steam is a source of creating heat and a decomposing agent at the same time. The roasting is performed in a short time, and with proper ore and pieces of the right size the result is very satisfactory. It is also a cheap process, and is applied for roasting gold quartz holding sulphurets, the amalgamation of which, without roasting, is defective. This kind of roasting would be also applicable as a preparatory for amalgamating silver ores.

In most instances with silver ores a most perfect chemical change is a condition on which the result of extracting the silver depends, and for this purpose the ore must be pulverized, in order to effect a perfect contact between ore particles, gases, and other substances which are mixed with the ore for certain purposes.

The finer the ore is pulverized the more effectual will be the intended purpose of roasting; but there are so many disadvantages connected with finely pulverized ore, that it is preferable to pulverize the ore as coarse as possible without interfering with a good final result. In the first place, the waste of ore in crushing increases with the fineness on account of dusting; the expense increases also, because less ore will be pulverized, and a still greater waste will occur during the roasting. It requires more extensive dust chambers than coarse crushing, but considering the very variable nature of ores, and also the mode of extraction, the fineness must also vary. Generally in pulverizing roasting ore for pan amalgamation, No. 40 wire sieve (40 holes to the running inch) is used; also, No. 35. Working the ore raw in pans without grinding, using the vitriol and salt, No. 60 has been applied with good success, but the crushing was performed dry, in order to avoid the loss of pulp carried off by water, although a sufficient number of settling tanks would retain almost all of the sloam; but the handling of such fine mud is very inconvenient, and by no means a clean job. Ore intended for lixiviation can be pulverized coarser than for pan amalgamation. Sieve No. 30 will probably answer in most cases; even No. 25 is used successfully (in Mexico).

The roasting of pulverized ore is now performed principally in continuous self-discharging furnaces, or in mechanical furnaces by charges, whereby stirring by hands is avoided. The old reverberatory have fallen into disuse, except in remote localities, or they are employed for roasting concentrated gold sulphurets. The continuous roasting furnaces are of diverse construction; but all agree in one important point, that the pulverized ore is conveyed by mechanical means from the pulverizing apparatus direct into the furnace, and discharge therefrom in the same proportion as fed by the batteries or other pulverizing contrivances.

Other mechanical furnaces, not continuous, are revolving cylinders of different descriptions, all so arranged as to be charged at once with from three to seven tons, and the whole discharged after six or eight hours roasting. The third class is represented by the old reverberatory furnaces, which in almost all cases give a satisfactory result, if properly treated. All have a horizontal hearth on which the ore is spread; all have at one end the fire-place, and the flue at the other, connected with the chimney. Between the flue and fire-place there are often two or three hearths, or one above the other. In accordance with the intended mode of extraction, the ore is either roasted with an addition of charcoal powder, whereby the silver is reduced to a metallic state—a procedure of no practical use—or the ore is subjected to an oxidizing roasting, with the principal object of driving out arsenic, antimony or sulphur, converting at the same time the silver into a sulphate (Ziervogel's process); or a chloridizing roasting is effected, that is, roasting with salt.

The roasting of gold sulphurets, and especially of silver ores, is of great importance, because ores of this class which require roasting, predominate considerably. This part of metallurgy deserves special attention, so much more as it represents the foundation of subsequent manipulations in extracting precious metals, either by amalgamation or lixivation. The oxidizing roasting is applicable only on gold-bearing iron sulphurets of the best quality, and never on silver ores, except sometimes in the beginning of roasting, then for the Ziervogel's process and for smelting purposes.

A Chloridizing Roasting in Reverberatory Furnaces.

In order to chloridize the silver, an addition of common salt is indispensable. The salt furnishes chlorine for that purpose, and is decomposed by sulphuric acid. The sulphuric acid is created by the

decomposition of sulphurets present in the ore.   It follows that if
silver ore is to be roasted successfully with salt, there must be a certain percentage of sulphurets in it; otherwise no sulphuric acid can
be obtained, and consequently no chlorination, or at least only an imperfect one, can be effected.

The time when the salt should be introduced, whether at the same
time with the ore, or after the greatest part of the sulphur and arsenic
has been expelled, is a question.   Many manipulators consider it
essential, especially with base ore, if rich in antimony and arsenic, to
commence the roasting without salt and to introduce it at the end of
the operation, one hour or half an hour before discharge.   The idea
is that arsenic and antimony, as chlorides, volatilize and influence
the silver also to become volatile, causing then a considerable loss.
Although this is true, it seems, however, that too much importance is
attributed to this behavior of the two former metals, because the antimony commences to escape at a comparatively low temperature, at
which little of the salt would decompose; and on the other hand, if no
salt is present, the silver volatilizes with the antimony anyhow; this
can be observed if the ore is rich in both metals as a pink colored
coating on the handles of the hoes.   An objection to the introduction of salt at the end of an operation is the difficulty of a perfect mixing.   In many works in Mexico, for instance, the salt is used
in a very coarse state, and then not only that during the last hour of
chlorination the mixing is imperfect, but the time is too short to decompose or volatilize the coarse particles, and after the charge is
drawn out a great deal of the salt remains undecomposed.   Another
objection to the charging of salt at the end of roasting is, that very
often little lumps of ore are formed in the beginning of roasting and
remain so during the whole operation, sometimes increasing in size.
If then the salt is added, this, of course, cannot enter the inside of
the lumps and the chlorination cannot be perfect; whereas, if the
salt had been introduced at the commencement, it would have been

diffused also in the lumps and the silver therein chloridized.   But in case there is so much lead in the ore that it commences to bake at an increased heat, then it is advisable to roast without salt and add it in the last hour of the roasting.

The most perfect mixing of salt and ore is doubtless effected in a battery.   The objection that choking of the sieves occurs if salt is used, seems to be groundless, unless charged in a moist condition. Almost all mills are compelled to dry the ore on a platform near the battery, or otherwise.   When the ore is spread, the necessary percentage of salt in a coarse state is scattered equally over the ore, dried and crushed together.   Some use half of the salt in the battery and add the other half to the ore in the furnace before discharging, but the benefit of this procedure is not evident.   It is, however, different with the mechanical furnaces, where the ore is conveyed by screws for some distance to the furnace.   If the salt is continuously added by a mechanical arrangement to the continuous feed of ore in the required proportion, when the ore enters the screw, the mixing of salt and ore as it moves forward under the screws is quite sufficient, but in this case the salt has to be dried and pulverized separately from the ore.   A very good arrangement for drying salt and ore is the revolving cylinder.   Before introducing the ore into the furnace the latter must be gradually heated up, which may take ten to fifteen hours.   When nearly red hot, a charge of dry ore, mixed with salt, is brought on the hearth through the roof and spread out equally by means of a hoe.   The fire is kept up moderately, but sufficient flame must be seen over the ore.   The draft is lessened by the damper, and the ore stirred diligently, but not continually.   The intervals, however, must be short.

In case the ore contains so much lead that it commences to bake at a little increased heat, this must be kept dark red, and a more frequent stirring is required for the first two or three hours.   By degrees the ore becomes red hot and the burning of the sulphur in presence of a great amount of sulphurets is quite lively. One part of the sulphur, by the action of oxygen, is converted into sulphuric acid and combines with the metals, deprived of their sulphur or arsenic, to sulphates. The period of the formation of sulphates is very important and requires some time before it is finished. If there is a large amount of sulphurets in the ore, the burning of the sulphur creates so much heat that the feeding of the fire must be stopped almost entirely for an hour or two, but must be resumed again as soon as it is perceived that the ore commences to cool. The workman stirs the ore, with a hoe or an iron rake, back and forward across the hearth, moving it from the bridge toward the flue and back. The formation of sulphates still continues with disengagement of sulphurous gas. The ore at the bridge is more exposed to heat than that on the opposite side, and the roaster is obliged to change the ore by raking it together into a long heap extending from the bridge toward the flue—not in the middle, but nearer the working door. By means of a shovel, six inches by twelve, on a long (12-foot) iron handle, the roaster takes the ore from near the bridge and transfers it toward the flue, putting it behind the ridge of ore until he reaches the middle of the furnace. He then takes the other end of the ridge and moves it toward the bridge. After this the stirring is continued in the usual way. As often as the roaster stops stirring, he should draw furrows across the hearth from twelve to fifteen inches apart, before he closes the door. The surface of the ore assumes then a wave-like shape, by which the area of the surface is more than doubled and the heat and oxygen have so much more exposed ore to act upon. This is important enough to be strictly observed. The sulphurets (combination of sulphur and metal), being now converted into sulphates (sulphuric acid and metal oxide) react now on the salt, decompose it at an increased heat, and set the chlorine free; a mutual exchange takes place.

Sulphate of Lead.

Sulphate of lead changes partly into chloride of lead, and partly it remains as a sulphate. This fact is of great importance for the subsequent pan amalgamation. All chloride of lead will be amalgamated and depreciates the bullion, while the sulphate does not amalgamate. According to an analysis of Mr. Stetefeldt's of roasted ore (in the Stetefeldt furnace) from Ontario, Utah, all lead that was in the ore proved to be in the state of a sulphate. This occurs also in reverberatory furnaces. For the lixiviation process the result is immaterial, because the sulphate, as well as the chloride of lead, will be dissolved in the solving solution and carried out with the silver together. At Plomosas, Mexico, the ore is roasted in reverberatory furnaces, and almost all the lead remains as a sulphate after roasting. The bullion from the lixiviation contained about 500 parts of lead in 1,000, while that from pan amalgamation is free of lead.

It seems that a high roof, 27 or 30 inches above the roasting floor, is more favorable for the formation of sulphate of lead than 20 inches or less. In the former case, there is more of the radiating heat that acts on the ore; in the last, the ore comes too much in contact with the flame. The chloride of lead volatilizes partly, and coming in contact with air, loses one part of its chlorine and is reduced to a combination of oxy-chloride of lead. Sulphate of iron and sulphate of copper change also into chlorides. The copper chloride becomes volatile, colors the flame blue, emits chlorine gas and forms subchloride of copper. The chlorine, set free, decomposes the sulphurets and sulphates of silver, and creates chloride of silver. If, during the operation, lumps are formed, in case the ore was not dry enough or too much heat was applied in the beginning, they must be crushed to powder by a hammer-like iron instrument with a long handle. As soon as the chlorination begins, after three or four hours, a different smell, that of chlorine, will be observed. White fumes arise, and gases and vapors are evolved, consisting of sulphurous acid, chlorine, hydrochloric acid gas, chloride of sulphur, of iron and of copper.

The ore increases now in volume, and assumes a wooly condition. Another hour's roasting will now finish the chlorination. This last hour's stirring requires a light red heat, in order to destroy as much as possible of the base metal chlorides. If there is a great percentage of copper and other base metals in the ore, the roasting may require more time, in order to decompose the chlorides and sulphates, the presence of which consumes too much iron, and during amalgamation in barrels, increases the heat to such a degree as to cause an injurious division of the mercury into small particles and scum. The base metal chlorides are reduced by the iron and also amalgamated.

The changing of the cooler portion near the flue with the hotter part at the bridge must be repeated two or three times during the roasting process. When finished, after five or six hours, the ore is drawn out and discharged through the discharge-hole in the bottom. White fumes and gases are still arising; and if left in a heap for several hours, the chlorination is still going on, and may gain 5 to 6 per cent. of chlorination.

The hoe is made of $\frac{1}{2}$-inch wrought iron, 6 inches high and eight inches wide. The rod or handle must be fifteen feet long at least. This would render the instrument heavy

Fig. 1. and tiresome to handle; it is therefore preferable to use a piece of gas-pipe, welding it together with the rod represented in Fig. 1.

Necessary Amount of Sulphurets.

In times when the barrel amalgamation was yet practiced in Freiberg (Saxony), long experience showed that a large amount of iron sulphurets was necessary, in order to decompose the amount of salt required for the chlorination. One hundred parts of the ore were mixed with 150 parts of borax glass, 100 parts of common glass, and one part of resin. This mixture, melted in an essay crucible, gave a button of matte (sulphide of iron), the weight of which was from 25 to 30 per cent. of the original weight of the ore. If less matte was obtained, the ore was considered too poor in sulphurets, and more pyrites had to be added.

There is not much silver ore found in the State of Nevada, which would give 25 per cent. of matte on the average; and as there are no pyrites to be obtained for this purpose, the ore must be roasted as it is. When starting the first amalgamation works in Nevada, I found from six to eight per cent. of sulphurets (different kinds) in the Comstock ore, which, after roasting, contained 88 per cent. of its silver converted into a chloride. The ore from the Rising Star Mine (Idaho) had not over 8 or 10 per cent. of sulphurets, still there was 91 per cent. of chloride of silver found after roasting. It is, however, very probable that from silver ores containing a great deal of calc-spar, or heavy spar, a less satisfactory result might be obtained by chloridizing roasting, if no more than 6 per cent. of sulphurets should occur in them. Some copriferous ores, especially if other base metals are present, and no sulphur (or very little), will give sometimes a good chloridizing roasting, without any addition of green vitriol or other sulphur combinations.

In treating ores entirely free from, or with a very small percentage of sulphurets, the want of sulphuric acid must be remedied by adding another substance. A cheap material of this kind is found in the green vitriol or copperas (sulphate of iron), of which $1\frac{1}{2}$ to 3 per cent. is added when 8 to 10 per cent. of salt is used. The copperas is first calcined, in order to drive out its water of crystalization, by a gentle heat, and from the calcined article, not the crystalized, is taken the above percentage. This sulphate acts then on the salt the same as if it were created in roasting. The copperas is also added to arsenical ores free from sulphurets. But the percentage of green vitriol to be added depends also on the nature of the gangue. If there is a great deal of lime in the ore it takes up sulphuric acid, forming sulphate of lime, remaining in this condition through the process of roasting without being decomposed further. For this reason calcerous ore requires as much more geeen vitriol or iron pyrites as is necessary to transform all lime into a sulphate. Silica or quartz, if abundant, in the presence of steam, decomposes some of the salt when red hot, forming silicate of soda and hydrochloric acid, the importance of which is shown by the fact that gaseous hydrochloric acid, in contact with metallic silver, unites with it to a chloride. It behaves in a like manner with sulphurets and arsenides, of which the most are decomposed, forming chlorides, while sulphur and arsesic escape combined with hydrogen.

Amount of Salt to be used.

Ores containing from 80 to 100 ounces of silver per ton should be mixed with 10 per cent. of salt. This is about the quantity considered necessary in the amalgamation works of Europe. Rich ore is often roasted with 20 per cent. of salt. If all the chlorine of the salt could be transferred to the silver, an insignificant amount of salt only would be required for ores containing 100 ounces of silver—not more than $3\frac{1}{2}$ pounds to the ton; but in consequence of the different ways in which the chlorine decomposes and unites with base metals and gases, the escape of chlorine from the surface of the ore without coming in contact with the silver, etc., a great deal more of the salt must be applied.

The usual amount of salt used for ores of the above value, is from 120 to 140 pounds per ton of ore; that is, from 6 to 7 per cent. It is not advisable to take less than 6 per cent., even if the ore be poorer. There are instances, however, where 91 per cent. of silver has been obtained by amalgamation from ores which were roasted with only 5 per cent. of salt, and even less. There was no natural chloride of silver in the ore when treated with 5 per cent. (Rising Star ore, Idaho.)

As the salt is not at all decomposed before the formation of sulphates commences, or only to a very small extent, it is also in this respect immaterial whether the salt is charged at once with the ore, or whether it is introduced two hours later, unless the ore is of such a nature as would bake easily on a little increase of heat. In other cases, however, it is obvious that, taking only 6 per cent. of salt, and employing only one man at a furnace, a perfect mixing in a short time, as ought to be done if the salt is charged after the sulphur is burned off, cannot be expected, and consequently a defective result will follow. It is therefore under such circumstances important to have the salt and ore introduced at the same time.

But a point of importance is the time when the salt should be added, if other objects are in view. If the salt is added together with the ore, or after the sulphur is expelled and sulphates are formed, in every instance the base metals will take up their share of the chlorine, and therefore more salt will be required. But as the most of the chlorides are volatile. the salt is the means of getting rid of a great deal of the metals during the roasting, which in some instances is not very desirable. For instance, if a great deal of antimony and copper is in the ore, more or less chloride of silver will escape; sometime, however, only a small percentage. Treating the ore with salt from the beginning, or adding it two hours after the beginning, the result is the same.

A different result is obtained if the salt is added after all the base metals are desulphurized and oxidized. Some base metals, as anti-

mony and arsenic, will be volatilized and thus gotten rid of, but not
in so large a proportion as if chloridized.   Iron and copper remain
entirely in the ore, while both are volatile as chlorides.   The roasting
must be continued at a light red heat till all sulphates are decomposed
and the metals oxidized.   Applying the salt after the dead-roasting,
the effect differs from the above so far, that the base metal oxides
now are not chloridized, or only to a small extent, while the silver
alone (some of which appears to be changed to a metallic state, the
most, however, remaining as a sulphate) will be chloridized.   But in
order to effect this chlorination, from 1 to 2 per cent. of green vitriol
must be added in order to accomplish the decomposition of all the
salt.   The copper is lost with the tailings unless smelted, or extracted
by diluted sulphuric acid.

Condition of Metals after Roasting.

The silver from silver sulphurets, native silver and other combinations, is turned principally into a chloride, but there may be found to
a small extent arsenate and antimonate of silver.   Copper from sulphurets and carbonates becomes an oxide and a small part of it is
converted into chloride and sub-chloride, but the latter will be found
in much larger proportion in case the roaster fails to increase the heat
enough to decompose the chloride, a considerable part of the copper-
chloride votalizes coloring the flame blue.   The lead from Galena is
turned into sulphate of lead, and in presence of antimony, partly also
into antimonate of lead, also in basic-chloride, of which a part volatilizes likewise.   Antimony volatilizes during the roasting to a great
extent as antimonious acid, another part combines with other metal
oxides to antimonates.   Chloride of antimony is also formed and is
also volatile.   Zinc, from zinc blend, is reduced to oxide, but partly
also transformed into a volatile chloride.   Iron and arsenic from
pyrites become oxidized, a small part of the iron remains as arsenate
and chloride.

Quartz and silicates remain unchanged. Calc and heavy spar, brown and fluor spar are turned into sulphates.

Permanent Stirring not Essential.

In roasting the ore with salt, a continual stirring to the end of the process is not a necessary condition for obtaining a good result. This depends partly on the time and partly on the nature of the ore. As long as the ore is not uniformly heated, a diligent stirring is important. The ore in the corners is too often neglected while the sulphur is burning, and the exposure of a fresh surface to the oxygen of the air requires also constant work; but as soon as the smell of the chlorine is perceptible, the stirring can be carried on at intervals of from eight to ten minutes. The chlorine which is evolved in the mass has better opportunity to act on the metals than if constantly stirred, whereby more chlorine escapes up the chimney without producing any effect. This was proved by a comparison of the work of two furnaces. A revolving furnace had a speed great enough to let the ore drop constantly through the flame and air, while the common furnace was managed by only one man, and stirred at intervals. Mr. Atwood found 15 per cent. less chloride of silver in the roasted ore from the revolving furnace. The blame is not with the revolving furnace, but with the speed. It proves, however, that being constantly exposed to the air, the chlorine escapes with less effect than in the common furnace, where the ore is allowed to rest for ten or fifteen minutes, and the evolved chlorine, being in contact with the particles while passing through the mass, is permitted to form combinations. O'Hara's mechanical furnace, in which the ore is comparatively but little stirred, gave 91 to 94 per cent. of chloride of silver. A mixture of ore, sawdust and salt, formed into bricks and calcined, showed the silver as a chloride through the whole mass,

where, as a matter of course, the inside did not come into direct contact with air.   Constant shoveling is necessary with ore of such a
nature, as it would bake if not stirred.

Signs of a good Chloridizing Roasting.

A good chloridizing roasting should give over 90 per cent. of the
silver converted into chloride of silver, and show as little as possible
of base metal chlorides.   To ascertain the amount of chloride of silver at the end of the roasting, it is necessary to make two assays.
Several pounds of ore are taken from different parts of the furnace,
well mixed and sifted.   From this, two one-half ounce assays are
weighed out, and one (No. 1) prepared for the fire assay as usual.
The other half ounce (No. 2) is introduced carefully into a small filter
in a glass funnel.   The filtering paper must project about one inch
above the ore.   A solution of hyposulphite of soda is then poured
over the ore in the filter, and this is continued as long as a precipitate
is obtained on adding one or two drops of a solution of sulphide of
sodium to the filtered liquid.   This is best tried in a glass tube.   If,
after filtering the addition of sulphide of sodium to the lixivium
does not produce a precipitate, or only a very slight one, so that the
liquid assumes only a little darker color, without losing its transpa-
rency, the assay is then leached with warm water and the filter taken
off, placed into a porcelain or sheetiron dish, dried, and the filter
paper burned above the sample, or with the sample in the muffle at a
low heat.   The sample, with the ashes of the filter, is fluxed like the
other half ounce, and both crucibles placed into the assay furnace.

The operation takes less time if the hyposulphite solution is used
in a warm state, but not boiling.   All chloride of silver, and also sulphate of silver, if present, is dissolved by the hyposulphite and carried
off, besides the base metal chlorides.   The two assays, when ready,
are compared, and the difference shows the silver which was convert-

ed into a chloride. For instance, if No. 1 assayed 83 ounces per ton and No. 2 from the filter 4 ounces, the difference, 79, is that part which became chloridized. That is,

$$83{:}79 = 100{:}x = 95 \text{ per cent.}$$

If there should be gold in the ore, this must be subtracted from both assays, because, although the amount of gold would be equal in both assays, the chlorination result, as it should be, must come out higher after the gold is subtracted.

The filtering operation, if properly attended to, takes one hour's time or more, and if a great deal of base metals is present it takes also more hyposulphite salt.

Hyposulphite of soda is a crystalized salt, of which five or eight ounces may be dissolved in a quart of water, warm or cold. This is the solving solution. It dissolves the chloride of silver much quicker than a common salt solution and is more convenient then to operate with ammonia. If the ore sample in the filter is treated with the hyposulphite solution, it takes in the beginning a great deal of soluble chlorides and sulphates, if present, but after a while always less and less; and in order to find out when all chlorides are extracted, a few drops of sulphide of sodium or of calcium, as before described, are added, and if no precipitation is observed, the leaching is finished.

For sulphide of sodium take five or six ounces of soda, or more, if a large crucible is at hand, melt it at a good heat, and, when liquid, introduce just as much sulphur (brimstone) at intervals in small pieces, giving time for the boiling up to subside. Pour out on an iron plate and dissolve in water. It takes several hours before the solution appears perfectly clear, showing a yellow color above the black sediment. When drawn off it is ready for use. Much better than sulphide of sodium is sulphide of calcium. It is easily made by boiling in an enameled iron kettle 3 parts of fresh burned lime and $1\frac{1}{2}$

parts flower of sulphur with water, enough to keep it thinly boiling for an hour. When cold the clear dark red liquid can be drawn off.

Toward the end of the roasting very little, if any, sulphate of silver will be found in the ore; but if a small percentage of it should remain, it may, for the purpose of amalgamation or extraction, be considered equal to the chloride of silver; for, as soon as it dissolves in water, it becomes a chloride, precipitated by the salt, of which a part is always yet found undecomposed in the ore. To obtain a general idea of the amount of soluble base metals, chlorides and sulphates, it is sufficient to put a small sample of about half an ounce on the filter as before, and to leach it with hot water. The leach obtained is tried again with the sulphide of sodium. A thick precipitate shows that a large amount of soluble chlorides is in the roasted ore. If a reaction of copper is expressly desired, ammonia should be used in place of the sulphide of sodium. In presence of much iron the precipitate will appear brown. This precipitate must smell strongly of ammonia. If copper is present, a clear blue liquid will be seen above the iron precipitate after some time; or the whole may be brought on a filter to separate the liquid from the precipitate.

To Regulate Chloridizing Roasting.

The main points which have to be considered at a chloridizing roasting are: The time of roasting, the temperature, the amount of salt and the loss of silver, all with reference to the percentage of chloride of silver. The first thing necessary to know, is the amount of silver contained in the ore—for this purpose a reliable average sample must be taken, including the salt if already mixed; this sample to be No. 1. It must be taken either while it is charged in the furnace, or at the battery every 10 or 15 minutes while it drops from the sieves. The roasting should be carried on at a moderate heat, as described (page 21.)

After three or four hours, when the ore assumes a wooly condition, other samples are taken from different parts of the furnace, mixed together, and the assay sample therefrom marked No. 2. One hour later a third sample is taken, No. 3; half an hour after this, No. 4, and before the discharge, or during the discharge, No. 5. The temperature of the last hour should be a moderate light red heat. In case there is a double or treble hearth furnace, one sample more should be taken from the hearth next to the finishing hearth, just before moving the charge over to the finishing hearth, but as there are three charges on three hearths, the samples have to be taken of one charge as it progresses from one hearth to the other.

While the next charge is treated in the same way, but a higher heat applied, the samples already taken must be assayed. In order to find the progress of chlorination, the assay of No. 1 is not needed, it serves only to ascertain the loss of silver during the roasting. One assay of No. 1, however, would not be sufficient, because the weight of the ore changes. For instance: the iron sulphuret (pyrites) loses its sulphur, which is replaced by the lighter oxygen, and weighs then 33 per cent. less, but in taking up chlorine, in place of the oxygen, it gets much heavier than it was when a sulphuret, gaining 35 per cent., but a great deal of the chloride of iron volatilizes. For this reason two assays of No. 1 are weighed out, one to be assayed in the usual way, to find out the amount of silver per ton, and the other half ounce is subjected to roasting under the muffle, in a roasting cup for one hour and a half or two, at about the same heat as in the furnace, then weighed out and noted how many grains were lost or gained, as compared with the original 240 grains.

The other samples are assayed to ascertain how much of the silver was chloridized.

If, then, all the assay samples showed a progress in chlorination, so that No. 5 was the highest, but still not satisfactory, it indicates that either the time was too short, or the heat too low, or perhaps not enough salt; these assays show also the relative loss of silver.

While these assays were made, the next charge, as before mentioned, was to be conducted at a somewhat increased heat, and samples taken as before. If now, for instance, the last two samples should show the same percentage of chlorination, it were evident that the roasting was carried on too long. In case the chlorination were still too low, and the last one the highest, then the roasting should be tried one hour longer—eventually more heat or more salt, but always controlled by assays. Such investigation may take two or three days, or more.

To ascertain the loss of silver, an average sample is taken when the roasting is finished, and for an assay, weigh out, not half an ounce, but so many grains as there were found after roasting of No. 1 under the muffle, and this weight considered as half an ounce. If then, for instance, the first assay of No. 1 should give 83 ounces per ton, and that of the roasted ore 78, that is five ounces less, the loss of silver by roasting would be $\frac{5 \times 100}{83} = 6$ per cent.

The loss can be considerably diminished by a proper arrangement of sufficient dust chambers, in which also the volatile metals could be condensed.

Means of Destroying Base Metal Chlorides.

It is very difficult to get rid of all the base chlorides. They are formed under the action of chlorine and hydrochloric acid. The most of the metal chlorides are volatile, and a part is carried off through the chimney. Another part of the chlorides gives off some of its chlorine, whereby sulphates, undecomposed sulphrets, antimonates and arsenates are chloridized. Chlorides which are disposed to transfer chlorine to other metals in combination with sulphur or arsenic, are: the proto chloride of iron and of copper, the chlorides of zinc, lead and cobalt. When in this way the most of the metals are chloridized, the base metals, principally iron and copper, are losing their chlorine gradually, being first converted into sub-chlorides, and then into oxides. The roasting for this purpose must continue with increased heat, even when the chlorination of the silver is finished. At an increased heat, the base metal chlorides lose their chlorine, while the chloride of silver remains undecomposed, unless a very high temperature should be applied. This process requires a long time, consequently also more fuel. The decomposition of these chlorides is greatly assisted by the use of 5 to 6 per cent. of carbonate of lime in a pulverized condition. Lime does not attack the chloride of silver, but it is not advisable to take too much of it, as it would interfere to some degree with the amalgamation. The pulverized lime rock must be charged toward the end of the roasting. First, two per cent. is introduced by means of a scoop, the whole well mixed, and then examined either with sulphide of sodium, or in the following way:

A small portion of the roasted ore is taken in a porcelain cup or glass, and mixed with some water by means of a piece of iron with a clean metallic surface. If the iron appears coated red with copper, some more lime must be added. In place of iron—especially if no copper, but some other base me al is present—some quicksilver is mixed with the sample. In the presence of base metal chlorides, the quicksilver is coated immediately with a black skin.

When endeavoring to expel the base metals by heat, the loss of silver, in presence of much antimony, lead and copper, should be investigated very carefully. Under certain circumstances it is not uncommon to find a loss of even 50 per cent. of the silver, if the chloridizing roasting is carried on at a high heat for a great length of time. The loss increases with the duration of roasting and with the degree of temperature. When such ore is under treatment, it is necessary to take samples during the roasting, and to examine the same

3

for the amount of chloride of silver, and also for its loss, and to stop roasting when the highest percentage of chloride of silver is obtained without reference to the condition of base metals.

Steam Decomposes Base Metal Chlorides.

The formation of base metal chlorides can be avoided by a proper but more expensive roasting. It requires, first, an oxidizing roasting, with or without the application of steam. This roasting must continue until all the metals are desulphurized and converted into oxides. When this is accomplished, salt and green vitriol are added, and the roasting continued until all the silver is chloridized.

There is also a very good way of getting out a good deal of the base chlorides of the ore before the silver is amalgamated or extracted, by leaching the ore with hot water.

Application of Steam in Roasting.

The application of steam in roasting is advantageous, for the reason that hydrochloric acid is created by the decomposition of chlorides, which in turn decomposes the sulphurets. The hydrogen decomposes also the chloride of silver, which, upon being reduced to metallic condition by its affinity for chlorine, in turn decomposes the hydrochloric acid. The silver may thus change repeatedly from a metallic condition to a chloride, while the base metal chlorides are reduced to oxides, and in that state do not interfere with the amalgamation or precipitation. The application of steam, however, requires a great deal more fuel during the roasting. Taking the moisture of the fuel into consideration, there is no roasting done without steam, although with a limited quantity.

Repeated experiments in California, Nevada and Mexico, applying steam through the bridge of the furnace, divided into four or five jets, never proved to be of such an advantage as to be continued, or permanently introduced where rebellious ores are treated.

Chloridizing Roasting of Silver Ore Containing Lead.

Lead has a bad influence in amalgamation, lixiviation, and in the roasting itself, causing a baking of the ore at the slightest undue rising of the temperature. The chloride of lead amalgamates easily, especially in iron pans. Ores with 8 to 15 per cent. of lead still allow of a successful roasting. A part of the formed chloride of lead escapes in gaseous form, another part is reduced by degrees to oxy-chloride of lead. This latter combination goes mostly into the amalgam. If there is more lead in the ore than 15 per cent., it gives sometimes, according to its nature, as much as 85 per cent. of silver, and the retorted amalgam is submitted to cupellation in order to separate the lead. If, as already stated (page 25), the lead is transformed into a sulphate, it does not amalgamate. This explains the reason why some bullion of plumbiferous silver ores is free of lead, while other bullion again contains sometimes 600 parts of lead in a thousand, according to the condition of the lead in the roasted ore. As yet no investigation has been made how to direct the roasting in order to convert all lead into a sulphate. Carbonate of lead as well as galena, form, under certain circumstances, sulphate, and very little of chloride of lead ; sometimes it is the reverse.

In Hungary (Offenbanya), black copper, containing, besides the silver, 10 per cent. of lead, was subjected to a chloridizing roasting. The pulverized copper is mixed with 12 per cent. of salt, 1 per cent. of green vitriol, and 3 per cent. of saltpetre. The saltpetre oxidizes the lead to a sulphate, which in the subsequent barrel amalgamation does not enter the amalgam. Here saltpetre is added for the purpose of oxidizing the lead to a sulphate in presence of salt. It is possible that an addition of 50 or 60 lbs. of peroxide of manganese to a ton of roasting ore may have a great influence in this respect; also the construction of the furnace is not immaterial (page 62.)

Difference in Roasting Ore for Pan Amalgamation as Compared with that for other Modes of Extraction.

The roasting of silver ores, if imperfect, will give a better result by amalgamating in an iron pan than in wooden barrels, or by lixiviation. This is due to the better decomposition of undecomposed sulphurets under the grinding muller. The roasting for pan amalgamation is therefore less delicate. However, when once at work, it is always better to do the roasting properly; but it is not necessary to sift the ore after roasting in order to separate the lumps from the mass, as is done with the barrel amalgamation, except to prevent nails from coming into the pan. The formation of such lumps, however, must be avoided as much as possible. Imperfect roasting in the presence of base metals gives in pans always a low fineness of bullion, as well as in the lixiviation process.

Roasting Charges.

The charges of ore in reverberatory furnaces in Europe are generally limited to five or six hundred pounds. This quantity can be handled very conveniently and all the theoretical details strictly carried out. The result of chlorination is very perfect ; but, for all this, the economy here in the United States is different with the different circumstances. The roasting charges here are generally 1000 to 2000 pounds, and if the chlorination comes up to 90 or 95 per cent., the operation can be considered very satisfactory.

Roasting of Silver Ores Containing Gold and Copper, to be Treated after Roasting by Chlorine Gas and Lixiviation.

It is not absolutely necessary for this process to have the ore roasted with salt, but it has been found that, on account of different earths, an addition of 1 or 2 per cent. of salt produces a better result. This process extracts copper, gold and silver, each of which is obtained separately; but it makes a difference in roasting, whether the copper is intended to be saved or not, as in many localities the copper is at present of no value, or the old iron for precipitation is too expensive.

The ore is crushed dry through a sieve of forty holes to the running inch. Some ore allows also thirty holes to the inch. In case the ore contains so much clay or talc that no leaching is admissible, the ore is crushed wet, separated from slime, and dried. Eight hundred to one thousand pounds are charged, and, according to the quality of the ore, the heat raised quickly or slowly to a bright red heat. This is an oxidizing roasting, consequently much stirring is required. Ores with but few sulphurets appear sufficiently well roasted after three hours; other ores, containing an abundance of sulphurets, take from five to six hours and more before all sulphur, arsenic and antimony are expelled; but the arsenates and antimonates formed by the two last are not volatile. When desulphurized, 1 or 2 per cent. of salt is thrown in the furnace and mixed with the ore as intimately as possible. Three-quarters of an hour after the addition of the salt, and with only a moderate heat, the roasting is finished. Ores, not rich in sulphurets, may be mixed with salt when charged.

If it is intended to extract the copper also, this must be transformed into chloride of copper. To accomplish this, two things must be observed—first, no oxide of copper should be formed during the roasting; and second, more salt must be used. Stechiometrically, each pound of copper requires 1.84 pounds of salt to form a chloride, provided all chlorine is taken up by the copper; but as this is not the case, as a great deal of chlorine is also absorbed by other metals, etc., it follows that at least two pounds of salt, if not more, must be taken for each pound of copper in the ore. For most localities such a quantity of salt could not be used on account of the difficulty in obtaining it. It may be mentioned here that if the brine remaining after the copper has been precipitated by iron, should be condensed by evaporation, exposing it to the heat of the sun, which might be practicable on the Pacific Coast in the dry season, the condensed salt, consisting of chloride of iron, could be added to the ore as a chloridizer, whereby a considerable percentage of salt would be saved.

In roasting with salt with reference to extracting the copper, the ore is first roasted for itself at a low temperature, so as not to decompose the sulphates by a bright red heat, but long enough to decompose all sulphurets. When this is accomplished, the salt is introduced and the roasting finished in one-half to three-quarters of an hour thereafter.

The process of chloridizing the ore with chlorine gas after an oxidizing roasting, for the extraction of gold and silver, and eventually the copper is now with much better result carried out by O. Hofmann's process, (page 122); but the roasting must be a chloridizing one, with the required amount of salt, (page 28.)

Chloridizing Roasting of Silver Ores Containing Gold, without Subsequent Treatment by Chlorine Gas.

Generally the ore containing silver and gold is roasted with salt, converting thereby the silver into a chloride, while the gold remains in a metallic condition. This mode of roasting is quite satisfactory for the subsequent amalgamation in pans. But if those metals are intended to be extracted by a solving process, where no amalgamation takes place, the gold also must be converted into a chloride while roasting. Roasting ores in which gold and silver is present with salt, chloride of gold is formed, according to Plattner; but before the ore becomes red hot, the gold loses a part of its chlorine, is reduced to a sub-chloride, and, at a little higher degree of heat, to metallic gold.

To form chloride of gold by way of roasting, a better result is obtained in the furnace if the ore is roasted first without salt till the smell of sulphur is no longer perceptible; and then, after it has cooled down to a low temperature, the salt is added, and the whole stirred for some time. A suitable form for a furnace would be a long hearth or furnace, altered in such a way that the second hearth should be ten twelve inches below the first; the arch of the first hearth, however, should continue in a straight line. By this means the space is widened and the temperature brought down to the proper degree. The ore is charged on the first hearth near the bridge, and roasted in the usual way oxidizing the sulphurets. When this is effected, the ore is shifted over in the lower furnace, and the upper charged again. As soon as the roasted ore assumes a dark red heat, the salt is introduced and raked for two or three hours. According to Roeszner, a combination of gold oxide of soda and chloride of sodium ($Au^2 O^3 Na Cl$) is formed. It is not soluble in water, and but slightly in salt solution, and cannot be amalgamated; but it is soluble in hyposulphite of soda or lime. V. Lill and others consider the gold in the state of a sub-chloride ($Au Cl.$) Hot water cannot be used for the purpose of leaching out base metals, as the chloride of gold would be decomposed.

Examples of Local Roastings.

Roasting of Silver Ores in Freiberg (Saxony.)

The amalgamation process, and consequently this kind of roasting, was given up long ago at Freiberg; but the method of roasting as performed there is nevertheless very interesting. The ore subjected to roasting consisted of silver glance, brittle silver ore, ruby silver, metallic silver, fahl ore, bournonite, zincblende, sulphide of antimony, iron, copper, and nickel pyrites, and of gangue, quartz, calc, brown, heavy, and fluor spar. It contained from sixty to one hundred ounces of silver per ton.

The dry crushed ore was first spread on a platform; on this a layer of damp ore, from the wet concentration, was laid, and then 10 per cent. of salt. This order was repeated from six to eight times.

The stratified mass was mixed thoroughly by means of shovels and a coarse sieve. This mixture contained from 9 to 10 per cent. of moisture. For this reason, after a charge of 450 to 500 pounds was introduced through a hole in the roof, the fire was kept very low, in order to try it at a dark red heat, and the ore was diligently raked by two men, working alternately.   As soon as the decrepitation of the salt ceased, the ore was ridged from the bridge toward the flue, through the middle of the furnace, and the formed lumps broken up by iron hammers attached to long handles. After this was done, the heat was increased, whereby the ore, under constant stirring, assumed a red hot condition, and the sulphur commenced to burn quite lively. This stage was reached in two hours from the beginning. The desulphurization commences with the burning of the sulphur, creating a temperature sufficiently high to continue the roasting without fuel for some time.   Fumes are evolved, consisting of steam, antimony, sulphurous acid, arsenic, etc. This desulphurization takes again two hours, the workmen all the time raking and changing the hotter part of the ore at the bridge with the cooler at the flue. The temperature is now raised to a light red heat; the ore increases in volume, emitting chlorides of metals, chlorine, hydrochloric acid, etc. The formation of the chlorides progresses rapidly, and is finished in three-quarters of an hour. The charge is then drawn out. A too long roasting would not give an equally good result, as some silver might be decomposed to the metallic state, which is not so readily amalgamated as the chloride.

Roasting of Argentiferous Copper Ores. At Arivaca (Arizona) the silver ores from the Heinzelmann mine consisted principally of silver copper glance(stromeyerite) with 51 per cent. of silver; fahl ore with from 2 to 15 per cent. silver, containing also some quicksilver; zinc-blende, galena and some decomposed argentiferous copper ores. On an average the ore contained from $150 to $200 per ton, and from 10 to 15 per cent. of copper. After crushing, the ore was spread on a platform covered with 8 per cent. of salt, and mixed thoroughly by means of shovels. Eight hundred pounds of it, as a charge, were introduced through the roof of the furnace (which was constructed entirely of adobe), spread on the hearth, and, at a dark red heat, stirred for two hours, at the end of which time the flame was colored—intense greenish-blue, and considerable fumes were emitted. The raking continued for an hour and a half more at an increased heat, and during this time the ore was moved three times from the bridge to the flue and back.

A sample taken from the furnace at this time, put on a canvas filter, wet with salt solution and leached with a hot concentrated solution of salt, gave a clear liquid, which, diluted with water, showed a strong white precipitate of chloride of silver, mixed with antimony and lead; but the quicksilver, treated with the same sample of ore and water, was cut and blackened to a high degree. For this reason, from 5 to 6 per cent. of pulverized lime was thrown into the furnace by means of a scoop, as much as possible over the whole surface of the ore, and then raked and stirred diligently in order to finish the mixing in the shortest time. After four hours from the beginning, the temperature was raised to a light red heat for half an hour, and the roasting was finished. There were yet a great many base metal chlorides in the ore, but as metallic copper was used in the barrels, the silver turned out always over 900 fine. The loss of silver was 12.5 to 13 per cent.

Roasting of Copper Matte. In smelting argentiferous copper ores, the process is sometimes regulated to produce a sulphide of copper, containing silver and base metals, as antimony, arsenic, zinc, iron, etc. This sulphide of copper, or copper matte, was roasted formerly and smelted again to produce black copper; that is, impure metallic copper. For the purpose of extracting the silver therefrom, the copper was melted together with a certain percentage of lead, and the latter, with the silver, extracted by liquation and cupelled. The remaining copper contained still some silver and lead, and the process was a very lengthy one before finished. To avoid the liquation, the copper matte was treated by amalgamation, and the silver extracted at once. For this purpose the matte was crushed and sifted, and the coarse part ground.

Of this powdered matte, 300 pounds are charged in a double furnace, of which the upper hearth prepares the ore by a moderate roasting, while the lower one finishes the operation at a higher temperature. In each of the hearth-departments the matte is treated for two hours and a half. Silver and the other metals are first converted into sulphates, and then mostly decomposed to oxides; but the silver remains for the greatest part in metallic condition. The matte is drawn out, mixed with 8 per cent. of salt and 12 per cent. of lime, and with salt water into a paste, which is allowed to rest for twelve or fourteen hours. The paste is then dried, powdered between rollers, and again roasted two hours and a half. During this process, samples are taken and mixed with water and a few drops of mercury. If this appears coated bluish, it proves the presence of metallic salts, and some more lime must be added; if after this the quicksilver remains perfectly white,—parting, however, in many minute globules,— it proves that too much of the lime was used, and in this case some of the first roasted matte is added.

The purpose of wetting the roasted ore, as above described, is the formation of chloride of silver. As there are always sulphates of the metals present after the first roasting, they decompose the salt, and the chlorine acts on the metallic silver. This process is not perfectly finished, and therefore the second roasting.

The roasting of copper matte is also performed differently from the above method; that the temperature in the upper hearth is kept so low as is sufficient to drive out the sulphur. Although the heat is very low, the stirring of the red hot matte is important; the more diligent the stirring is, so much less lumps will be formed. This roasting on the upper hearth may require five hours, after which time it is drawn on the lower hearth, and, under constant stirring, roasted for two hours more at a moderate heat; then the temperature is increased, and the roasting continued for two or three hours more, till no evolution of fumes is perceived. A sample taken from the furnace and leached with water, should color the filtered solution hardly perceptible blue. The copper is now mostly turned into oxide; also the iron. The silver is converted principally into a sulphate, and if there is some lead, zinc, arsenic, and antimony in the matte, then these metals are found to be sulphates, arsonates and antimonates. In small quantities these metals are not injurious; but lead and antimony, if too much of it, are troublesome.

In order to chloridize the silver, about five per cent. of salt is now introduced into the furnace, after the ore is cooled down a little, or a part of it is taken out and mixed with the salt on the floor, charged back in the furnace and mixed with the balance of the ore. Increasing the temperature, the salt soon commences to volatilize and chloridizes directly, and partly also in contact with some sulphates. During this chloridizing roasting the temperature is kept always moderate till to the end. If there is only copper, iron and silver in the matte, all these metals are partly chloridized, the first two assisting the chlorination of silver; but if the roasting before the addition of salt was imperfect, leaving too much of the sulphates undecomposed, then a great deal of the copper volatilizes as chloride of copper, and causes at the same time also a loss of silver. If the matte contains also lead, zinc, antimony and arsenic, these metals volatilize as chlorides and cause a still greater loss in silver and copper.

Roasting of Black Copper. The black copper obtained from smelting (in Schmoellnitz, Hungary) contains from 110 to 150 ounces of silver per ton, and 85 to 89 per cent. of copper.   To pulverize this it must be made red hot in a reverberatory furnace and crushed while red hot.   The powder must be sifted and then ground fine.   The pulverized metal is then mixed with 7 to 9 per cent. of salt, and roasted in the usual way for from six to six and a half hours, in charges of 400 pounds each.   No green vitriol is added for the purpose of decomposing the salt; and as there is not more than from $\frac{1}{2}$ to 1 per cent. of sulphur in the black copper, the salt decomposes through direct action on the copper.   First, chloride and sub-chloride of copper are formed.   The copper chloride transfers chlorine to the metallic silver, and is reduced to a sub-chloride.

In other places iron pyrites is added to the black copper, by which the chlorination is promoted.   At Oravitza (Banat) 5 per cent. of iron pyrites and 12 per cent. salt are used, roasting twelve hours. The loss of silver is 7 per cent., and of copper 3 per cent., during the roasting.   The expense of roasting is $7.30 per ton.

Roasting of Silver Ores for the Patera Process.   The ores treated by Patera's process (page 128) at Joachimsthal are remarkable for the numerous mineral species occurring in the ore.   Among these may be mentioned silver, lead, different compounds of copper, bismuth, iron, uranium, nickel, cobalt, etc., sulphur, arsenic and antimony.   Before the introduction of Patera's process, the extraction of silver, on account of so many base metals, was very difficult.   The success of Patera was not so much in adopting hyposulphites, as proposed by Percy and Hauch, but in his modification of the roasting process, by which only the silver was converted into a chloride.

The pulverized ore is placed in the furnace, in charges of from four to five hundred pounds.   First quite a moderate heat is applied, and gradually increased, but not so much as to induce clotting.   As soon as the ore appears red hot, steam is admitted, with about four pounds pressure to the square inch.   As the steam consumes heat, more fuel must be used to keep up a red heat. The ore must be constantly raked during the whole period of this roasting. It takes about four hours to finish this process, after which the ore is drawn and permitted to cool. The iron appears now as an oxide—also the copper ; some sulphate of copper is also present, and the silver is principally in the state of a sulphate.

The oxidized ore is now ground finer, mixed with from 5 to 12 per cent. of common salt, and, at the same time, with 2 to 3 per cent. of calcined green vitriol. This mixture is spread upon the hearth in the furnace, and subjected to a second, now chloridizing, roasting. It takes about an hour before a red heat is reached. Steam is then introduced, as above, under continuous stirring. The fire is gradually increased, and the roasting finished within from five to eight hours, according to the value of the ore. There are condensing chambers for catching volatile metals and ore dust. They are of importance if rich ore is treated, and without this contrivance several per cent. of silver would be lost. When finished, the ore is drawn and allowed to lie undisturbed for some time, after it has been moistened. In this condition the chlorination continues. The application of steam causes nearly twice the consumption of fuel, but it has been shown that, by the steam, hydrochloric acid is formed, whereby arsenic and antimony are expelled, and at the same time the chlorination of the silver greatly favored. Over one ton of coal and one-half cord of wood are consumed for each ton of ore roasted in this way.

Roasting for Augustin's Process. To this process principally matte is subjected. The process requires a chloridizing roasting. The ore, if not itself rich in sulphurets, is mixed with iron pyrites, slag, lime, etc., and smelted. The molten sulphide of iron takes up the silver and deposits itself below the slag on the bottom of the furnace. The silver is thus separated from the earthy part and concentrated in the matte. Argentiferous copper ores are likewise smelted for the purpose of obtaining argentiferous copper matte. The matte is then finely pulverized, and 400 pounds of it are introduced into the furnace. The roasting goes on now in the usual way, by starting with a moderate temperature, gradually increasing it, exposing every portion of the ore to the intense heat by frequent stirring, etc. At the end of eight hours the roasting is generally completed, the matte looks dark and earthy, and no fumes of sulphurous acid can be perceived. The roasted stuff is now drawn out and permitted to cool. After this the matte powder is ground finer, sifted or bolted, and given back to the furnace.

Four hundred pounds are mixed with 5 per cent. of salt. Sulphates are present, but oxides of metals predominate in the mixture. The formation of the chlorides commences immediately, and the roasting is concluded after one or two hours. The temperature in this second roasting is kept low, as the smelting of the chloride of silver must be prevented—for when melted the chloride of silver dissolves with more difficulty in a salt solution. In other places, after the first roasting, the mass is not taken out and re-ground, but, when finished, only a portion is drawn, mixed with from 1 to 6 per cent of salt, according to the purity of the matte, charged again, mixed with the balance of ore in the furnace, and roasted for one-half to three-quarters of an hour.

Roasting of Silver Ores in a Long Furnace at San Marcial, Sonora, Mexico. For the purpose of roasting silver ores for the Solving and Precipitation Process, there are several long furnaces (thirty feet long) built by Mr. O. Hofmann, in Sonora. Using long furnaces is found a great economy in every respect. A great advantage results also from moving the ore, at intervals of from one to three hours, from one hearth to the other. By this means it is impossible, even with careless roasters, to find raw ore in the finished charge, as would happen under such circumstances if the corners of single roasting furnaces were not very carefully attended to during the roasting.

The furnace at San Marcial is sixty feet long. It is a level hearth sixty feet long, representing six furnaces, each ten feet long, parted only by the projecting wall inside.

There are six working doors on one side, and, on account of the length, an auxiliary fire-place is placed at the back side, before the last two hearths. Each hearth contains 800 pounds of ore, and is attended by two Yaque Indians at a time, stirring alternately. These twelve Indians perform all the work about the furnace, carry the wood from the adjoining yard, split what is too thick, carry out the ashes, etc. The ore on the first hearth, nearest to the fire-place, has always a light red heat, which decreases with the distance from the bridge, so that the fifth hearth appears quite dark if not assisted by the second fire-place. After stirring one hour on all six hearths, the charge from the first is drawn out through a door in the rear, on a large, smooth platform, and immediately spread by means of shovels in a layer one or one-and-a-half inches thick, so as to have it cool enough for transportation to the sifting apparatus after the lapse of one hour. As soon as the hearth is cleared, the ore on the second hearth is moved over to the first, then that on the third to the second, and so on, till the sixth hearth remains empty and is charged through the funnel in the roof with a new charge. It will be seen that 800 pounds of ore are drawn out every hour, and that each charge is exposed to the fire for six hours. It is thus evident that, being moved six times from one hearth to the other, the ore arrives perfectly prepared to the finishing heat. After roasting and sifting, the ore is amalgamated in pans, but as it contains some carbonate and sulphuret of lead the amalgam is charged with base metals, so much that refining by cupellation is necessary. From 8 to 10 per cent. of salt is added and mixed with the ore before it is charged. Preparations are made now to introduce the solving and precipitation process, if successful on that kind of ore.

According to an analysis made by Mr. Graff, the roasted ore from single furnaces (treating $100 ore) contained 5 per cent. less of chloride of silver than that from a long one. Long roasting furnaces are especially adapted for roasting sulphurets containing gold. Concentrated sulphurets, or ore containing an abundance of sulphurets, allow the use of a very long furnace, with only one fire-place, on account of the heat created by the burning sulphur.

The roasting expenses at San Marcial, with the furnace sixty feet . long, were as follows:

24 men, day and night, at 50 cents	$12 00
2 cords of wood at $3	6 00
8 per cent. of salt=1,536 ℔s. at 2 c.	30 72
Repairs, etc.	3 00
Total expenses on 9.8 tons	$51 72

or $5.27 per ton.

A furnace thirty feet long, with the same kind of laborers, 800-pound charges, drawing every two hours,—that is, 4.8 tons in 24 hours, shows the following expenses:

8 roasters at 50c.	$ 4 00
1½ cords of wood at $3.	4 50
8 per cent. of salt at 2½c.	19 00
Other expenses	2 00
Total	$29 70

or $6.18 per ton.

B.—Oxidizing Roasting.

The purpose of the oxidizing roasting is either to expel volatile substances which are combined with the metals (as sulphur or arsenic), or to expel volatile metals which are considered obnoxious to further treatment of silver ores (as antimony, lead, zinc, etc). The oxygen has a large share in this transaction, and combines with the volatile substances, as well as with the metals. Some of the combinations with the oxygen become volatile—as, for instance, sulphurous, arsenous and antimonious acids, lead and zinc oxides, etc. Other combinations again are not volatile, as the formed sulphates, arsenates and anti-monates. Some of these latter compounds can be disengaged by an increased heat, as the sulphates of iron and of copper, whereby the sulphuric acid escapes, while the remaining metal turns into an oxide. Others cannot be decomposed by an increased heat, or an increased heat is considered injurious for other reasons; and in this case such combinations may be decomposed by, an addition of charcoal powder, saw-dust, or the application of hydrogen. The sulphuric acid is reduced by the carbon to sulphurous acid, and goes off, and so also the arsenic and antimony. The carbon deprives the sulphate or arsenate of a part of its oxygen, and escapes as carbonic acid.

What Process Requires Oxidizing Roasting.

The oxidizing roasting is in use partly as a preparatory treatment for a chloridizing roasting. For Ziervogel's process alone an oxidizing roasting is in use. For amalgamation of silver ore it is not suitable; but it is important with the smelting processes, and also in extracting gold from gold ores—principally from sulphurets (iron pyrites). For this purpose long furnaces (Fig. 8) are the most suitable, but also double furnaces.

The main point in the roasting for Ziervogel's process is the creation of a sulphate of silver, and the oxidation of the base metals as far as possible. As the decomposition of sulphates of different metals depends on different degrees of temperature, such roasting appears of a very delicate nature. To this process principally argentiferous copper matte is subjected.

. *Roasting of Copper matte.* When pulverized until fine enough to pass through a sieve of thirty-three holes to the running inch, the mass is introduced into the furnace and spread out by means of rakes. The matte inclines much to clotting. For this reason a very moderate

4

temperature is applied, more for drying than for roasting. The matte
is left quiet for about fifteen minutes, after which the stirring is
commenced and continued without stopping for an hour. During
this time many lumps are formed, which the roaster tries to crush to
powder. Near the working door the stuff is exposed to a draft of
fresh air, in consequence of which the roasting on that place progresses more rapidly than it does further back. This makes a
shifting of the stuff necessary after one hour's roasting. The other
roaster now takes the rake and stirs the matte again for an hour,
doing the work precisely as the first roaster did. The roasters change in
this way every hour for from five and one-half to five and three-quarters
hours. This roasting is performed on the upper hearth of the double
furnace. Twenty-five pounds of coal dust are mixed with the matte,
causing an ignition and emission of gases, and the whole mass is
transferred to the lower hearth through a hole in the bottom. The
upper hearth is now charged with 500 pounds of matte anew.

The sulphur commences to burn after a raking of three-quarters of
an hour, and the mass increases in volume when the hearth is covered
about four inches with matte. During the roasting all metals are
converted into sulphates, of which, toward the end of the operation,
iron and cinc vitriol are decomposed, leaving those metals as oxides.
Copper, nickel and cobalt remain in the state of sulphates.

The lower hearth is in a light red hot condition when the matte
falls in from the upper hearth. To prevent the rapid burning of the
admixed coal dust, and the clotting of the mass, a vigorous stirring
for an hour, with closed dampers, is strictly observed. The stuff is
now shifted and then the damper opened. There now follows a sharp
oxidizing roasting, with free access of air, for one hour and a half.
By means of the air current, the roasting mass is cooled down so far
that it appears quite dark. To see the progress of the roasting, a
sample is taken out and examined, either on a porcelain dish or on a

filter with cold water. The leach must appear of a clear blue color, and an addition of salt solution must give some white precipitate, proving the beginning of the formation of sulphate of silver. If the filtrate shows a greenish-blue color, the presence of sulphate of iron is apparent, and in this case the oxidizing roasting must continue.

The purpose of the addition of coal dust is the reduction of sulphates to basic salts, whereby sulphurous acid is emitted. With the opening of the damper the oxidation progresses, the sulphate of iron is decomposed almost entirely, the sub-oxide of copper turns into oxide, and when the oxidizing roasting is finished, the mass contains mostly oxides, but also basic salts. There are copper, iron and zinc oxides, sulphates of copper and zinc, while the silver as yet consists principally of an undecomposed sulphide. The next stage of roasting at an increased temperature is the last one. It is directed toward the sulphatization of the silver and complete oxidation of the base metals. It takes two hours and a half to accomplish this result, under continuous raking and increasing the temperature to a light red heat. Samples are taken again as before, and examined in the same way. The leach must appear only of a bluish tint, and on adding salt solution, a heavy precipitate must fall, caused by chloride of silver. The whole roasting period on the lower hearth, as on the upper, takes from five and one-half to five and three-quarters hours.

The formation of the sulphate of silver in the last period at an increased heat, is due to the sulphuric acid in gaseous form, emanating from the sulphate of copper. It attacks the sulphide of silver, and combines with it to a sulphate. Ninety-two per cent. of the silver is extracted after this roasting. If in the last period the feeding with fuel should be carelessly performed, so as to give a smoky flame, some copper oxide will be reduced to sub-oxide, and this will precipitate metallic silver while leaching, causing a loss. If the roasting should not continue long enough, some sulphide remains undecomposed; and, on the other hand, if the roasting should last too long, a part of the sulphate of silver would be decomposed to metallic silver and could not be leached out. These circumstances show that this kind of roasting demands a great deal of attention, in order to obtain a perfect result. The temperature on the lower hearth in the beginning is 500 to 550 degrees Centigrade; it sinks then to 425, and rises again at the end of the operation to 770 degrees.

Roasting of Iron Sulphurets Containing Gold, and of Tellurides of Gold.

The gold is generally found in a free state as metallic gold. In this state it is easily extracted by proper amalgamation. Often, however, the gold is combined with other substances, so that amalgamation is of no avail unless the gold is set free by roasting. Iron pyrites and arsenical pyrites are the principal ores containing the gold in a condition unfit for direct amalgamation. Also telluride of gold must be subjected to roasting before amalgamation.

It is not decided whether the gold in sulphurets exists in a metallic condition, finely impregnated, or whether it is, like the iron of the pyrites, chemically combined with the sulphur. Some believe because after the free gold is extracted, and the sulphurets still finer pulverized, some more free gold appears, that for this reason, all the gold in the sulphuret can be set free if only a sufficiently fine grinding is effected. This conclusion, however, does not seem to be correct. The sulphurets may contain free gold, so minute that after each grinding some free gold is exposed, but for all this there may be a great deal of it chemically combined, which never can be set free by trituration. There are sulphurets in California assaying $1200, and more, per ton, and they may be ground to the very finest sloam, but no free gold, or only very little, will be visible under a good microscope. This seems to confirm the theory of a chemical combination much more, than the presence of some free particles of gold after a finer trituration, which proves only that free gold is in sulphurets, but not that there is no chemical combination besides the free gold. The roasting of the

Telluride of Gold

or of telluride of silver rich in gold (Petzite) is a most delicate operation. Not all tellurium combination with gold are losing the gold to a notable extent while roasting, but some do, and that to a considerable amount, up to 20 per cent., perhaps even more. The loss is no mechanical one occasioned by the draught of the furnace, but principally by volatilization. This occurs sometimes at a very low heat. As soon as the ore gets hot enough to show rising fumes, it is advisable to take a sample and assay it and not to increase the temperature until the result of the assay is known.

As an example in what singular condition the gold is in some combinations with tellurium is the Petzite, containing 24.80 per cent. of gold and 40.60 of silver. Heating a small piece of this mineral on a piece of charcoal or on the blade of a knife to the temperature of about 323 deg. 6', the melting point of lead, the bluish gray mineral turns suddenly yellow, without changing its shape. Under the microscope this yellow color is found to be due to innumerable perfect gold globules which cover the surface. The gold separates from the silver and tellurium in a molten condition, while the other two metals remain unchanged. The gold requires 1200° 6' to become liquid, and in this case it melts even below 323.

If salt is present during the roasting, the chloride of tellurium volatilizes voluminously, and it is possible that under this condition the tellurium causes the gold to volatilize likewise. The author, by his own experience—having 4 per cent. of salt in the ore on account of silver—found the loss of gold amounting to 8 per cent. before the ore was properly red hot. This was two hours after charging. Under

this circumstance the continuation of roasting was impossible ; the ore,
therefore, was drawn out and then chloridized with chlorine gas, and a
good result obtained; not, however, without lixiviating for a while
with chlorine water after the largest part of the gold was first extracted
after chlorination with gas.   Whether the above loss would have
occurred in such proportion if no salt had been added, cannot be stated.
It is indispensible, then, to have condensing chambers connected with
the roasting furnaces, if tellurides have to be treated, in order to con-
dense the volatilized metals.

In Transylvania (Hungary) the tellurium ores are subjected to
roasting, then smelted for matte, the matte granulated, treated with
diluted sulphuric acid, and the residue, containing gold, silver and
lead, fused with lead ores, and the resulting lead cupelled.   In Colorado the tellurides are also matted with other ores, the matte pulverized and roasted oxidizing after Zervogel's method to convert the
silver into a sulphate, which is leached out with water, then the residue
fused again to black copper, which contains the gold.

At Zalathna, Transylvania, the gold telluride is fused direct with
lead in cupell furnaces, but the tellurium in this case is lost.   If
matted with iron and copper sulphurets, and the silver extracted by
leaching after roasting, the gold is separated from the copper by diluted
sulphuric acid, producing marketable sulphate of copper.

The roasting of iron sulphurets is a very simple oxidizing process.
The main object is to effect a perfect desulphurization in order to set
free the gold.   Generally, no loss of gold is suffered during the roasting, neither with iron pyrites, nor with the arsenical pyrites, although
exposed to a considerable heat and for a long time (from 24 to 48
hours).   Many manipulators, however, who have had much experience
in roasting of gold sulphurets, complain of considerable loss of gold
with some kind of sulphurets, which they could not avoid in spite of

all experiments made with reference to heat and general treatment. It is, therefore, always advisable to investigate the loss of gold with new sulphurets by assay.

The simplest way to ascertain whether the loss occurs in roasting or otherwise—because the loss may happen also in the subsequent treatment and be wrongfully charged to roasting—is to weigh out one-half ounce of the well sampled and dried sulphurets before charging the furnace, then another half-ounce after roasting. But the sulphurets, after roasting, weigh a great deal less than before, having exchanged the heavier sulphur for the lighter oxygen, consequently the gold appears concentrated in the smaller weight of the ore. For this reason it would be improper to take half an ounce of the concentrated and compare it with that of the original not concentrated sample. In this case if half an ounce of the roasted ore were assayed, the number of ounces would come out higher, and the actual loss either covered, or the gold would seem to increase in the furnace. If salt is used in roasting sulphurets, it is generally charged in the last hour, and the sample should be taken just before the salt is added, and only so much for the intended half ounce, as found after roasting of half an ounce as described (page 34.)

The weighed sample of the raw sulphurets is introduced into No. 8 French crucible, in which a mixture of one ounce of soda and one ounce of litharge was previously placed, and then carefully mixed with a glass or iron rod. After this, a 20-penny nail is inserted into the mixture and the whole covered with about one ounce of borax. The other sample of roasted ore is also introduced into another crucible, having in addition to the soda and litharge mixture eight grains of powder of charcoal, well mixed and covered with about one-half ounce of borax in small pieces, no nail being required. Both crucibles are exposed to a good heat in a furnace for 15 minutes, the covers taken off, the nail with the cupell tongs carefully taken out, giving a gentle jerk before the nail is out of the crucible, the contents poured out and the buttons cupelled. Using the nail as above, the time which would have been consumed in roasting is saved and the result perfectly reliable ; but samples of arsenical pyrites, copper sulphurets or iron pyrites containing arsenical pyrites, must be subjected to roasting.

Iron sulphurets, well concentrated, carrying only an admixture of quartz, require only a perfect oxidizing roasting in order to give a satisfactory result in the subsequent chlorination ; but if lime, talc or clay is with the sulphurets, an addition of salt in small quantities, twenty to forty pounds to the ton, is necessary, otherwise the precipitation of gold is sometimes very troublesome, and as the presence of salt is not injurious it is always better to use some salt with the sulphurets. The quantity of salt must be increased if silver should be present. For the roasting of sulphurets, the reverberatory furnaces are almost exclusively in use, and of these again long ones, with two or three hearths, also, one hearth above the other. These last furnaces have the advantage that the floor is heated from below, and the burning of sulphurets on the upper hearth comes sooner in heat. There is also a better mixing of more and less ignited parts, in drawing the charge to the lower hearth, and the roasting can be performed by one man as well as in the long furnace, unless the sulphurets are of such a nature that the roasting is finished in a much shorter time than usually.

The roasting hearth is generally 10 feet by 10, or 12 by 12, receiving one ton of sulphurets. After the furnace has been heated up, the sulphurets are introduced and spread over the hearth, and the fire kept up lively enough to bring the sulphurets in a red hot condition. After this it takes very little fuel, as the burning is maintained by the sulphur itself. Nearly half of the sulphur is expelled by this low heat. On exposing a fresh surface of the mass by stirring, the burning of the sulphur with a bluish flame can be seen distinctly. The hoe is principally used for stirring. · It must be as light as possible, seven to eight feet long, if prepared to work from both sides of the furnace. The stirring is performed at intervals of ten to fifteen minutes, and whenever the circumstances permit two roasters to be employed, the time of roasting will be shortened. Oxidizing roasting requires more stirring than the chloridizing.

In proportion as the oxidation of the sulphurets draws nearer to the end, the temperature decreases, and it is then necessary to use more fuel to keep the mass at a good red heat. It takes from twenty to forty hours before the roasting of one charge in a single furnace may be considered finished. If, in throwing up sulphurets in the furnace by means of a shovel or hoe, many brilliant sparks appear, this denotes that the roasting is not finished, but must be continued till this appearance ceases.

In a long furnace, (Fig. 8) the hearth near the bridge is always kept at a bright heat. One man attends to the ore on the first hearth, and the other two or three hearths can be managed by a second. In moving the ore from one hearth to the other, or in drawing the charge from the finishing hearth, these two men assist each other. The finishing hearth receives the ore already desulphurized to a great extent, containing only a small part of undecomposed sulphurets, but more of sulphates. With a lively heat and active stirring at intervals, all base metals ought to be converted into oxides after ten or twelve hours. The heat must be increased to a bright red heat, but not to a white heat, else the gold particles would melt, which, with a good magnifying glass, can be easily detected after washing off the iron; the gold appears then in minute globules, the chlorination of which is more difficult.

Taking a sample of the sulphurets at the end of the roasting, and placing in a filter and leaching with a little water, the leached liquid will contain the soluble metals, and if, by addition of sulphide of soda

or of lime (page 33), a thick precipitation is observed, the roasting
should continue longer.   The mixing of sulphurets with 20 or 30 per
cent. of sand depreciates the value, consumes more wood in roasting,
requires more chlorine gas, more handling; but there is no evidence of
an advantage in so doing, unless a better result is obtained by having
more chlorine gas on less sulphurets, the volume of which is increased
by admixture of sand; but in this case it would answer better if the
chlorine gas (treating pure sulphurets) should be allowed to pass
through the same for an hour after it appeared on top of the charge.
There is no danger of backing or matting of pure iron sulphurets if
a light red heat is applied, unless the temperature should be raised to
a white heat, but this were beyond the limit of roasting heat.

Roasting Furnaces.

Roasting not only requires much care, but it is also an expensive
operation.   For this reason the choice of the right kind of furnaces
is of very great importance, and so much the more as a perfect and
economical extraction of silver depends principally on the result of
roasting.   The chloridizing roasting is known to be the most suitable
way for the subsequent extraction of silver in whatever way it may
be performed, by amalgamation or solving; consequently those furnaces in which the ore particles are exposed to the action of chlorine
and other chloridizing gases to the most advantage, must be considered
the best.   The old style of furnace was four to six feet wide and ten
feet long, and in them a small part of the ore was exposed to the
greatest heat near the bridge.   The gases evolved were carried along
by the draft, being in contact with the surface of the ore for a length
of ten feet while passing over it; but on account of the narrowness
of the hearth, the ore at the bridge had to be changed often with the
cooler part at the flue.

The next step in improvement of reverberatory furnaces was the
adoption of wider hearths, even wider than long.   The heat was more

uniform and the result better.   In both kinds of furnaces the chlorination of the metals depends principally on the chlorine developed
in the mass of the ore while passing through it; but once above the
surface, the chlorine and volatile chloride metals have less chance to
transmit their chlorine to the ore, and this only through the chlorination period.· During two or three hours of each charge, when desulphurization and sulphatization are going on, this must be performed
by the oxygen of the air, while, if chlorides were present from the
beginning, sulphurets, sulphates and oxides would have been partly
decomposed directly by the chlorine, whereby time and a certain
percentage of salt are saved.

In this respect a great advantage is gained by the introduction of
"long furnaces," in which a continual formation of chlorides on the
finishing hearth near the bridge is going on, volatile chlorides and free
chlorine being evolved, which, on their way to the flue, are constantly
in contact with the ore for a space of thirty or fifty feet in length.
These furnaces show a great economy in fuel, labor and salt, and the
roasted ore contains a better percentage of chloride of silver.

The reverberatory furnaces, although combining many valuable
properties, have the great disadvantage of requiring much labor to
perform the roasting, which, in this country, in remote localities,
amounts to four dollars and five dollars a day per man. In order to reduce this heavy expense, mechanical furnaces of various construction
have been introduced with great success, but often there are local circumstances which favor the adoption of reverberatory. A very important
improvement in the way of chloridizing roasting is found in the
Stetefeldt and some other furnaces, where all ore particles are involved
in chloridizing gases under very favorable circumstances. The roasting is cheap, and from twenty to thirty tons of ore are roasted in
twenty-four hours.

The roasting furnaces do not require a white heat, hence common
bricks can be used; but it is nevertheless advantageous

if the fire-place above the grates is built of fire bricks.
In new or unpopulated districts even unburned bricks or adobe
may be used; they stand just as well as burned bricks of the
same material, except in the floor of the furnace, which is worked
out in two or three months.   Hard bricks are the best material for
the hearth-floor, placed edgeways (four inches high) with as little clay
between as possible, and laid carefully and well fitting, so as to form
a level and smooth surface.   All parts exposed to heat must
be built with loam or clay, not with mortar.   Many masons have the
custom of laying three heights of bricks so that the eight-inch wall
is formed by two rows lengthways, and only the fourth height is put
crossways.   It is a quick work and may answer for buildings, but
should not be allowed with furnaces where the expansive heat must
be considered, especially in the fire-place.   Each alternate row of
bricks must be laid crossways to the preceding; also adjusting the
wall with the hammer, to make it perpendicular and square, after
several bricks are laid, is injurious.   The outside appearance of a
furnace is of minor importance, and the mason must, contrary to his
general idea, pay the most attention to the solid and particular work
inside.   The distance of the arch from the hearth is from twenty-six
to thirty inches in the highest point, not far from the bridge; in a long
furnace, however, the roof of the first hearth can be higher from the
floor by four to five inches, according to the length.   An eight-inch
thickness of the arch is sufficient, and the bricks laid with the eight-
inch side perpendicular form a more durable arch than one of twelve
inch thickness composed of eight and four-inch sides of the bricks.
The furnace must be secured against expansion by wall-braces of
cast-iron, tightened with iron rods from five to sixth-eighths of an inch in
diameter.   The rods placed over the length of the furnace are stronger
—one inch in diameter.   In place of wall-braces also wooden
posts, six by eight inches, are used, tied by iron rods on the top.   The
lower ends are generally put in the ground, but it is preferable to use

rods on both ends. In case of need, even the rods are replaced by timber. For the passage of the rods, square holes must be provided in the masonry; also for the escape of dampness such passages are necessary at different points, especially if the whole block consists of masonry. The floor of the hearth should be three feet and a half above the ground; if lower, it is inconvenient for the roaster.

There are two principal classes of furnaces—such as are managed by hand and such as employ machinery. For the first class, mostly reverberatory furnaces are in use. Of the second class, the most important are the cylindrical and vertical roasting furnaces, and of the last ones only the Stetefeldt furnace.

A. Roasting Furnaces Managed by Handwork.

Reverberatory Furnaces. Reverberatory furnace is the name applied to all horizontal hearth furnaces provided with grates and fireplace on one side, and a flue connected with a chimney on the other. The draft here is created by the chimney instead of by bellows, as in blast furnaces; therefore only such fuel is used which gives a flame, and consequently no charcoal, coke, or anthracite is serviceable unless in a gas reverberatory furnace, where gas (carbonic oxide) is produced from charcoal or other fuel—sometimes also by the aid of compressed air—and burned. The reverberatory roasting furnaces are constructed in various ways. There are single furnaces, with but one hearth, and double furnaces with two hearths, one above the other. Sometimes above the second hearth there is a third one for the purpose of drying the charge. Long furnaces are preferable.

A Single Roasting Furnace is represented by Fig. 2, showing the section, and Fig. 3 the ground plan. The bottom, *a*, or the hearth, is made of the hardest bricks, laid edgewise and as close as possible. Some masons lay the bricks flat. This mode is cheaper and quicker, but far inferior and less durable than the former way, and requires a

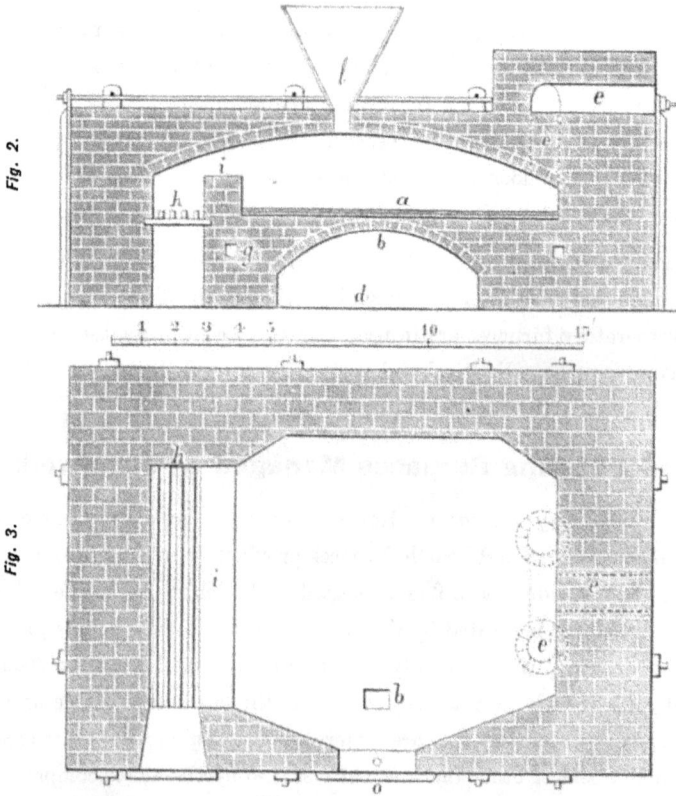

more carefully prepared foundation. The very best bricks must be selected for the hearth. *b* shows the discharge hole in front of the hearth. It is more convenient to draw the ore toward the front hole than to have a door for this purpose behind, but circumstances may decide for such discharge doors. The flue, *e*, is in connection with the flue-holes, *e*, in the arch, as indicated by dotted lines in Fig. 3, and is from nine to ten inches in diameter. The flue-holes in the arch have the advantage that no ore can enter when being stirred, as often happens when the flue commences at the hearth. The distance between arch and hearth near the bridge is twenty to twenty-one inches and near the flue only eight inches. The flue leads into the chimney in any suitable direction, either directly or through a dust chamber. Often the flue is led under the floor (when the chimney is at some distance from the furnaces), and is made wide enough to serve as a dust chamber—say two feet wide and three feet high, or wider if several furnaces are connected therewith. The chimney is from twenty to fifty feet high, and from one and one-half to three or four feet square in the clear. On the top of the chimney an iron cover, controlled by a chain, regulates the draught. This is practicable only when but one furnace is attached to the chimney, otherwise dampers must be provided for each furnace in the flue. The bridge, i, is much exposed to injury by fire on one side, and by raking on the other; it is therefore advantageous if the upper part, or the whole bridge, can be made in two or three parts and of some fire-proof stone—sandstone, granite, or some conglomerate, which does not burst when heated. The grates, h, are twelve to sixteen inches below the top of the bridge, eighteen inches wide, and from six to seven feet long. The space between the grate-bars is one-fourth to one-half of an inch.

In the roof, near to the bridge, is an opening four to five inches square, of cast iron, in connection with a funnel, l, of sheet iron. This funnel must be large enough to receive one charge of the ore. A slide keeps the ore in the funnel. The roof must be either eight inches thick, or the double length of the brick; that is, 16 inches. Under the hearth there is an arched space, d, into which the roasted ore is drawn through the discharge hole, b, either directly into an iron car or on an inclined floor, on which the ore slides from underneath the furnace. In front this space is shut up by brick work. For the purpose of easy drying it is well to have open some holes, g, for the escape of dampness. It is not necessary to build the block under the hearth solidly of bricks. The space inside is generally filled up with rubbish of bricks and stone, or it is still better to build the floor on arches.

The working door, *o*, is from twenty-five to thirty inches wide. In front of it is an iron roller for easier handling of the heavy tools. The door is eight to nine inches high. The cast iron door frame, for the fire-place, is from nine to twelve inches square. When completed, the furnace is tied by iron rods, *n*, both ways. The uprights are often wooden ones, six by six or five by eight inches. In place of the usual iron roller in front of the working door a far better contrivance, which allows the hoe easily to be directed to the required points, which is not the case with the roller bars, is the following, illustrated by the engraving, Fig. 4. The wheel, *a*, about $2\frac{1}{2}$ inches in diameter with $\frac{3}{4}$ inch journals, has half hollowed face to receive the stem of the hoe. This wheel rests in a frame, the sides of which are forked to receive the journals of the wheel, and ends in a pin, *e*, which is inserted in a corresponding hole of the doorframe, *d*, Fig. 5. This can easily be attached to doorframes already fixed on the furnace, from outside, by drilling two holes, and then to screw on a piece of flat iron in the shape as seen in Fig. 5. The hoe moves easily to and fro on the wheel at the same time with the turn of the fork to any required direction of the hearth. Fig. 4 represents one quarter of the natural size.

Fig. 4.   Fig. 5.

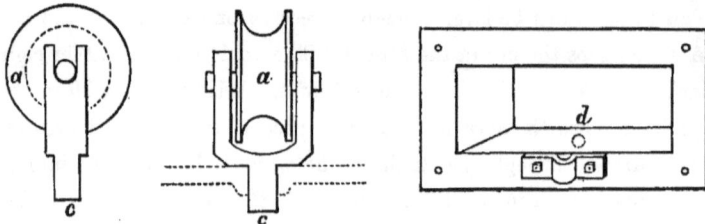

It is very important to dry the furnaces, when finished, with a very moderate fire for five or six days, day and night. Upon a slow, gradual drying, the durability of the arch depends. The furnace must be nearly red hot before the first charge of ore is introduced.

In building a roasting furnace where the locality is favorable to such an arrangement, it is a great advantage to lace the furnace so that the floor on which the roaster stands should be 10 or 15 feet higher than the cooling floor, wherefrom the ore can be carried to the amalgamating pans or to the leaching tubs, as represented by the outline sketch, Fig. 6.

Fig. 6.

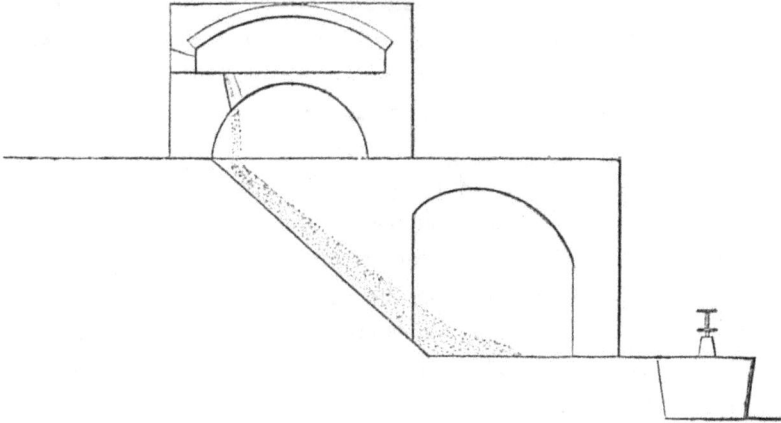

Fig. 7.

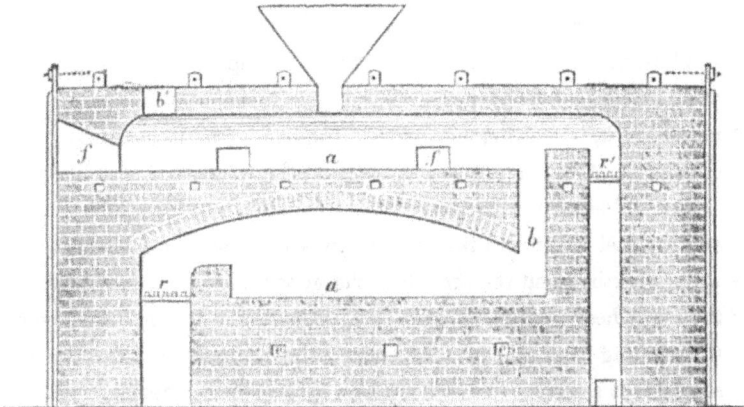

A Double Roasting Furnace is represented in Fig. 7, in longitudinal cross section. The lower hearth, *a*, is nine feet long and ten feet wide. The roof in the center is 28 inches, and at the flue and

5

bridge fourteen inches above the hearth. The fire-place, r, is twenty
inches wide, eight feet long, and twenty inches from the roof. The
flue, b, ascends to the upper hearth, c', the working door, o, is on the
back side. In case there should be required more heat than is obtainable from the lower hearth, there is an auxiliary fire-place, r'. The
flame goes through the flue, b', into dust chambers. These chambers
have cross partitions lengthways, by which the draught is forced to
take a longer way before it enters the chimney. From the upper
hearth the ore is drawn through the flue, b, to the lower hearth; e, e,
are canals for the escape of moisture, and $e' \, e'$ for the tie-rods.
The two hearths can be used separately if needed. This furnace,
although somewhat inconvenient for the roasters, has the advantage
of taking up less space, and as the ore drops from the drying hearth
on the second, and from there on the first hearth, it effects a good
mixing of the not uniformly heated ore. The auxiliary fire is useful
for silver ores, which seldom contain so much sulphur as to burn long
enough without the help of a nearer fire than that from the lower
hearth. In working concentrated sulphurets, the upper fire can be
dispensed with. Deetken, who has had long experience in roasting
sulphurets, considers this kind of furnaces superior to the long ones
and to all others he ever used. The arch in the middle is only eight
inches thick, but, being made level, the thickness increases towards
the sides. A few days after starting up, the whole roof is sufficiently
heated to keep the entire mass of the sulphurets red hot throughout,
and the fuel is better utilized; one ton always being drawn from the
upper hearth, when the preceding charge is finished. The charge on
the lower hearth requires, generally, from eight to ten hours before
the roasting is completed. In constructing a double furnace, in two
stores, it is essential to use a number of iron ties across the width of
the furnace between the lower and upper hearth; otherwise, on account
of the weight of the upper charged hearth, the side wall would inva-
riably give way in the course of a while. If used for the roasting of

Fig. 8 *Fig. 9*

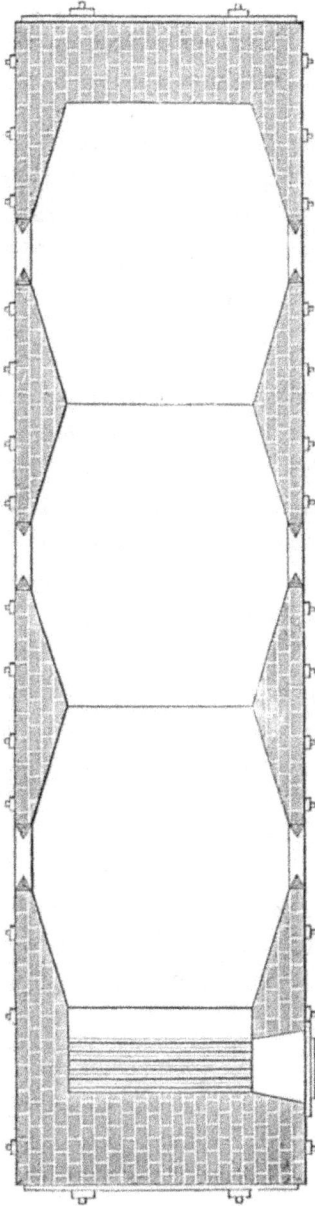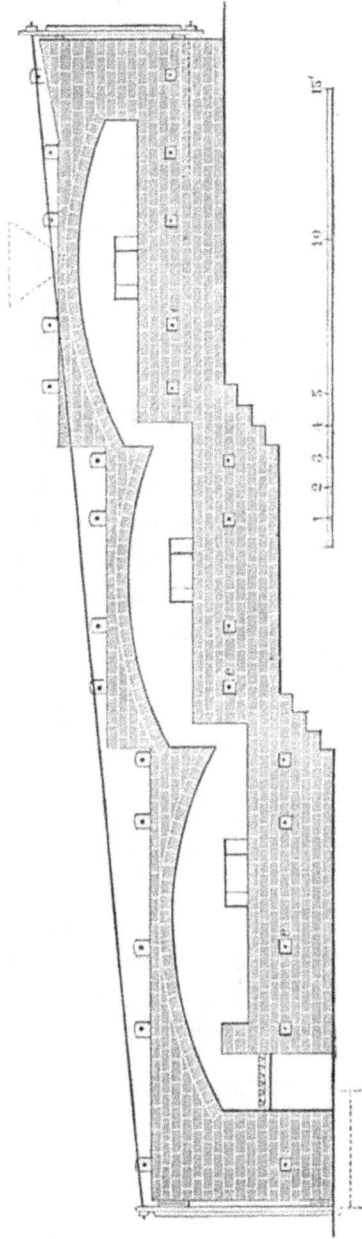

gold sulphurets, or such ores as do not require the auxiliary fire, the working doors can be placed at *m* and *n*, and, at *m*, a hoe can be advantageously used in drawing the charge to the lower hearth.

Long Roasting Furnaces. This kind of furnaces, as represented by Fig. 8, pg. 71, by a vertical section, and Fig. 9, by the ground plan, gives much satisfaction, as there is not only a great saving of fuel effected, but also a greater quantity of ore can be roasted than with a single furnace. It is a modification of the double furnace, and seems to be more convenient for the roasters. There are two men employed at a time, there being one ton and a half to two tons in the furnace. The hearths are either arranged horizontally, as the drawings show, or only the first one is level, the other two are inclined; this facilitates the shifting of the ore. Each hearth is ten feet long and twelve feet wide. After the first hearth there is a step of four to six inches, partly to divide the first from the others, but principally to effect a better mixing of the ore. The ore is fed on the last hearth, through the sheet iron funnel, spread equally on the hearth, and, according to its dampness or the quantity of sulphurets contained, stirred more or less from one and a half to two hours. As it is not only inconvenient, but impossible, to have a good stirring effected at a distance of twelve feet, which requires long and heavy tools, there are for this reason working doors on both sides of the furnace. The roaster uses hoes or rakes eight feet long, made partly of gas pipe, which are light and handy. The working doors are thirty inches wide. They must all be kept closed except when the ore is being raked, and then it is very proper to have half of the door closed (with a piece of sheet iron). Sufficient air comes in at the working door of the first hearth.

After one and a half to two hours the ore is removed to the middle hearth and spread equally over the whole surface. A new charge is introduced in place of the former. There is a higher heat on the middle hearth than on the last one. The treatment of the ore is here the same as before, being raked from time to time. After a lapse of one and a half to two hours the ore is moved again to the first hearth, in the same way as before. The ore is now exposed to a light red heat, by which the chlorination or oxidation must be finished in the same time as on the other hearth. It is necessary to change here the ore from the bridge toward the flue, and reverse once during the roasting. When the operation is finished, the roasted ore is drawn into iron cars below the furnace, through the opening in the floor. When all the ore has been removed, the charge on the second hearth is transferred to the first, from the third to the second, and from the funnel to the third hearth, and the process continued as before, so that a thousand pounds are drawn out every one and a half to two hours.

The bridge is fourteen inches high. For the purpose of admitting air or steam, a canal can be made in it. The fire-place is eighteen inches wide and eight to nine feet long, and fifteen inches below the top of the bridge. The ash-pit is made according to what seems more convenient, or as represented either in Fig. 8 or in Fig. 9. A deep ash-pit is more favorable for the preservation of the grates, as they are less heated. Each door is provided with an iron roller. A furnace of a similar description was in operation in La Dura (Mexico), roasting refractory silver ores for the chlorination process.

A furnace sixty feet in length, with six hearths, as built by Mr. Graff at the San Marcial, has the advantage of being capable of roasting from eight and one-half to twelve tons of ore in twenty-four hours, discharging every hour from eight hundred to one thousand pounds, according to the charge. In case ore is subjected to roasting which has not enough sulphur to create the required heat in burning, an additional smaller fire-place must be attached on one side, so as to bring the flame into the fourth hearth.

Muffle Furnaces.

A muffle furnace, as the name indicates, is a furnace constructed of clay and cast iron in such a way as to prevent the flame from coming inside of it. The fuel heats the mantle or muffle from the outside, so that the ore is not heated directly by the burning fuel, but by the glowing muffle. The muffle furnaces require, therefore, more fuel to obtain a certain degree of heat than ordinary reverberatory furnaces, where the flame comes into contact directly with the ore. The use of this furnace is limited, and applicable in cases where the air or the gases of the burning fuel are injurious, or where volatile substances from the ore should be condensed; as, for instance, sulphur, zinc, arsenic, etc. For roasting silver ores, these furnaces are not in use, but they were tried in California in different ways; also, for desulphurization, adding charcoal to the pulverized ore. The experiments, however, were not successful, as could have been anticipated.

Mechanical Furnaces Fed by Charges.

There is a great variety of mechanical furnaces whereby the costly stirring by hands is replaced by mechanical contrivances. On account of the very different styles of arrangement, the description of the treatment of the furnaces fed by charges, as well as those fed continuously, will be found, together with the description of the furnace itself, unless they agree in the main points, like the revolving cylinder furnaces. No mechanical furnace can be governed in every part of the roasting process with the same facility and precision as is possible in a reverberatory furnace with manual labor; but in the latter case the great difficulty in finding good reliable roasters, and the heavy expenses connected therewith, make a mechanical substitution very desirable. Amongst this class of furnaces are the revolving cylinders, through which the flame passes the most successfully. The fuel is here better utilized than in a reverberatory furnace; but, being twelve to fifteen feet long, the ore is more exposed to the heat at the fire-place than on the opposite side. To avoid this, Brickner uses a diaphragm, which does good service in renewing the surface of the ore; but, to bring the ore from the flue to the fire-place and back, it does not quite give entire satisfaction in this respect. M. B. Dodge made recently another arrangement with projecting bricks, the result of which is not yet known. A cylinder of a large diameter will heat the ore more uniformly than of a small one, both having the same length. Besides the cylinders there are furnaces, the hearth of which revolves, having a circular shape and stationary stirrers. Other furnaces have a stationary round hearth and revolving stirrers, etc.

Bruckner's Revolving Furnace. This furnace consists of a cylinder of boiler iron, lined with fire-bricks as tight as possible. This cylinder revolves between a fire-box and the flue on friction wheels. The flame passes from the fire-box direct into the cylinder, and from there, with all the gases that evolve during the roasting, into the dust chambers. Inside of the cylinder is the diaphragm, made of cast-iron pipes. It is set at an angle of about fifteen degrees to the axis. It has a diagonal position extending through the whole length of the furnace, and is intended to move the ore from one end to the other in order to expose the cooler ore from the flue to the stronger heat at the fire-box. In this way the stirring is performed by aid of the diaphragm and the revolution of the cylinder. The great drawback of unskilled or indifferent roasters is hereby avoided. What the roaster has to perform is so simple that any person of common sense can learn the treatment, after having been once regulated, in a few days. The expenses for labor and fuel are considerably less than with reverberatory furnaces; also, the time of roasting is generally shorter. If properly arranged, one man can attend two furnaces—that is, to fire up, charge and discharge. One charge is from one and a half to two tons of silver ore, and, according to the ore, from five to six per cent. of salt, which is

charged together with the ore, or towards the end of roasting. The salt should be ground fine and dry; if damp, it will form small balls, which remain so during the whole process. There is a manhole in the cylinder, through which the charge is introduced, and the same hole can serve also for the discharge, or to get the charge out quicker; in others there are also two manholes in place of one, one opposite the other. When the ore is charged, the door is closed and the cylinder allowed to revolve at the rate of one or two revolutions per minute, and the roaster can concentrate his attention to the one and principal task, viz: to keep up the right temperature. He can judge of the heat partly through an opening that is made expressly for this purpose opposite the flue, through which he should be able to take out a sample with a long handled scoop, and he can also see through the iron pipes of the diaphragm from outside, which pipes appear more or less red hot, according to the heat inside of the cylinder. Like in a reverberatory, so also in Bruckner's, very little fuel is used as long as there are sulphurets enough in the ore to keep up the heat by burning of the sulphur. The fire must be regulated accordingly and must be increased as soon as the temperature commences to sink. By degrees the heat is raised to a greater intensity. When the reaction between salt and sulphates begins, the ore assumes a woolly consistency, increasing in volume considerably. The disagreeable smell of sulphurous acid disappears and chlorine gas is evolved in its place; the temperature is now increased to a bright red heat for an hour, and the chlorination is considered finished. The motion of the cylinder is stopped for a moment, the manhole door opened, and the revolution of the furnace started with an increased speed of from 5 or 6 revolutions per minute.

In Colorado, where Bruckner's furnaces were successfully introduced in 1867, for roasting of silver ores and gold sulphurets, the ore is made to drop through a grate into a screw conveyor, and conveyed in an iron trough to the required place for further treatment.

This is a very convenient way to move the hot ore away from the furnace, and less expensive than if performed by hand labor. In order to obtain satisfactory results, the furnace must revolve slowly, about one revolution in one or two minutes.

Modified Brucker's Furnace. The difference between this modified cylinder furnace and Bruckner's, consists principally in the omission of the diaphragm, and is thereby reduced to a simple cylinder furnace. A large sized cylinder was built in the Pacific Foundry, San Francisco, capable of holding seven tons per charge.

In treating rebellious ores, this furnace gives very satisfactory results. The speed of revolving is best ascertained by experiments, trying first with one revolution per minute, then one in two minutes, and one in three. In roasting ore, the treatment in this furnace does not differ from the preceding.

B. Dodge's Revolving Furnace. This furnace differs from the former in shape, being six sided, in stead of round. The furnace is seven feet in diameter and sixteen feet long, calculated for a charge of seven tons. The gearing is so arranged as to alter the speed without altering the speed of the driving engine; the change of speed is necessary for the discharge, when the roasting is finished. The revolving of the furnace is effected by peculiar gearing. The bands encircling the furnace rest on a pair of grooved rollers. The shafts of one pair of these rollers are extended past the bearings, and on their ends are pinions, between which is placed a gear wheel. On the opposite end of the shaft, carrying this gear, is a crown wheel, and this being riveted, the center gear rotates the pinions both in the same direction, thus moving the rollers in the same direction, and rotating the cylinder, applying the power on both sides.

The six-sided cylinder is lined with bricks, forming also peculiar projections, calculated to move the ore from one end to the other. The charging is effected through two doors; and there are two other doors opposite, so that when all four are open at an increased speed, the discharge of the seven tons takes but little time. The doors are so arranged that they can be opened or shut with the greatest ease. The advantage of the six-sided shape of the cylinder is said to be in the fact that, as the furnace revolves, the ore, resting on the rising side, remains stationary, being held there by the corner, formed by the junction of the sides. At a certain point, however, the whole mass on that side drops down, and is completely turned. The outer surface of this body then remains in contact with the heat and flame during the time the side which is carrying it is raised up as the furnace rotates, when another sudden fall of ore takes place, and the operation is repeated. This of course can take place only after the ore assumes a woolly or spongy condition. Some other mechanical furnaces of less importance, are those of a circular hearth. Of this kind there is the

Revolving Hearth Furnace. The shape of this furnace is circular. There is an iron frame of from ten to twelve feet diameter, with sides fourteen inches high. The whole is lined with brick, the bottom four inches thick. The discharge opening is on the bottom, extending from the periphery toward the center, and is four inches wide and three feet three inches long. This opening is shut by an iron door, hung on hinges. It is not necessary to fill this space with brick, which would interfere with the easy opening; but the space, after the discharge of the ore, must be filled up with roasted ore, of which enough is always left in the furnace. The bottom is fixed to an upright shaft, four inches in diameter, provided with a spurwheel at the lower end to impart the motion. This ten or twelve-foot bottom is surrounded by a substantial ring wall, as close to the periphery of the bottom as possible. The bottom is then arched over with bricks, leaving the doors through

which the new shoes are introduced when the old ones wear out.
There is also a cast iron pipe through the center of the furnace, on
which the shoes are fastened and so arranged that one set plows the
ore against the center, the other set toward the periphery.   The pipe
is hollow and cooled by a continual stream of water.   There is also
a hole four to five inches square in the arch, in connection with a
funnel, through which the ore is charged into the furnace.   The
distance from the bottom to the center of the arch is thirty-one
inches.   The arch is connected on one side with the fire-place, six or
seven feet long and eighteen or twenty inches wide, and about ten
inches below the rim of the revolving hearth are the grates.   On the
opposite side is the connection with the flue.

   Such furnaces have the advantage that they carry the ore in a
circle, so that each part is equally exposed to the heat near the bridge
and to the cooler region near the flue.   While revolving, the funnel
is opened and the ore falls on the moving bottom, being spread in
passing under the stationary stirrers, which are of a plow shape.   The
roasting takes about the same time as in an ordinary furnace, but
requires less fuel, as the furnaces are cooled down by air, which enters
the common reverberatory furnace through the working door.   It is
important to have the horizontal shaft provided with two driving
wheels of diffent size, so that about one to two revolutions per
mieute can be obtained while roasting, and from six to eight revolu-
tians while discharging.   After the roasting is finished, the discharge
door on the bottom can be opened, while the hearth revolves slowly.
In this furnace it is an easy matter to arrange the plows in such a
way that they could be moved every second or third day toward the
periphery as much as they wear off.   In this way the side of the
hearth can be kept always clear from accumulation of the ore crust.

   A similar furnace is *Burton's Revolving Furnace*.   The hearth has
a low conical shape, the highest point being the center.   Above this

is the charging hole in the roof. The hearth is twelve feet in diameter, and takes one ton of tin ore at à charge. There is a cast-iron rake with three-inch long prismatic teeth, which are dovetailed and so constructed as to be easily replaced. The ore comes through a funnel in the center of the revolving hearth, and is spread by the stationary rake, the position of which is not radial but oblique. The hearth is fixed to a solid vertical shaft with gearing, by which a slow rotating motion is imparted to the hearth, so that only one revolution is made in forty minutes.

Parke's Roasting Furnace, with movable stirrers. This is a double furnace, one hearth above the other, with a common vertical shaft to which the stirrers are fastened. The hearth is twelve feet in diameter and rests on an arch, beneath which the rotating motion is transferred to the shaft by means of gearing. On one side of the lower hearth is the fire-place, whence the flame draws over the bridge into the furnace.

Opposite the bridge is an opening one foot wide and four feet long, through which the flame ascends to the upper hearth. Both of the hearths have two working openings, which are closed by cast-iron doors. From the upper hearth the flame draws through a flue into the chimney. The shaft goes through both hearths and the roof. There are two massive arms in both furnaces, with curbed spikes attached for the purpose of stirring the ore. In order to keep the shaft cool, it is hollow, and a few holes above the gear permit the cold air to draw through the shaft, whereby a constant cooling is effected. The upper end of the shaft runs in a cast-iron cross, fixed on the roof of the furnace.

After the ore on the lower hearth is drawn out through the dis charge hole at the bottom, the ore on the upper hearth, already desulphurized to a great extent, is raked toward a similar discharge hole, and then transferred to the lower department. The raw ore is charged through the roof into the upper part. By means of hoes the ore is spread on both hearths, before the shaft is allowed to revolve again.

Mechanical Furnaces with Continuous Feeding.

This class of furnaces has a great advantage over the preceding ones, inasmuch as there is still less manual labor connected with the whole operation of roasting, charging and discharging. From the crushing machinery the ore pulp is conveyed by screws, and, if necessary, by elevators to the furnace, and the roasted ore is discharged mechanically in the same proportion as it is fed. The arrangement of these furnaces differs in construction very much, as will be seen from the following description :

*The Howell Improved White Furnace.** This furnace consists of a tellescope-shaped iron cylinder, as represented in figure 10. The wider part of the furnace, where the flame enters, is lined with fire-bricks; the other narrower part of the cylinder is not lined at all, but stands the less intense heat very well; it receives during the operation a kind of ore-coating, which protects the iron. The cylinder has a slightly inclined position, and is of a very simple construction; the revolving arrangement is much improved and durable. The usual size is 40 inches in diameter and 24 feet long, capable of roasting 10 to 20 tons of ore. The largest cylinder is sixty inches by twenty-four feet, and calculated for thirty-five to forty-five tons. The forty inch furnace requires two horse power, and weighs ten tons. It is made in sections for transportation by wagons or mule backs.

The main fire is at the lower or discharge end of furnace. The flame passes over the ore receiving pit, directly into the cylinder, and

* This furnace is patented and made in the Pacific Foundry, San Francisco.

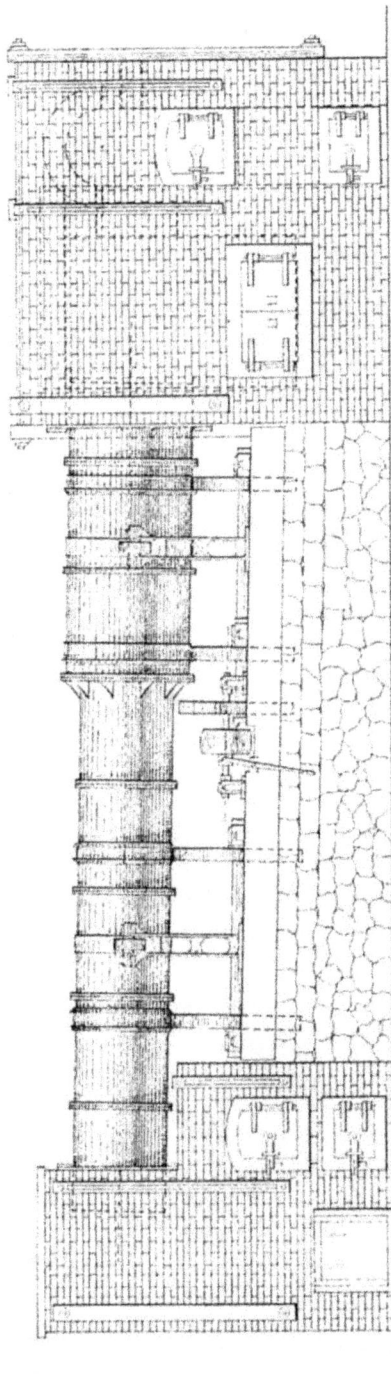

through it towards the flue; the ore on the contrary is fed on the opposite side, the upper end, and progresses against the flame by degrees into the higher heat. The ore is regularly fed directly from the battery or some other crushing apparatus, by means of a screw conveyor, either with salt from the battery, or the salt can be fed mechanically outside the battery by a salt-feeder, which can be easily regulated to feed the required percentage. In the lining of the cylinder there are brick projections about two feet apart, and in the upper part which is not lined with bricks, there are flanges to prevent the sliding of the ore, and to carry it up by the revolving furnace till it drops through the red hot air and gases, to be lifted again. This action is constantly repeated; the ore at each drop comes nearer to the discharge end and into a higher temperature, till it drops out into the ore-receiving pit, well roasted and chloridized to from 85 to 95 per cent. of the assay value.

By the constant showering of the ore and the draught, the dust of the ore becomes separated from the sandy part, and is carried out of the furnace towards the flue, before it can reach the necessary temperature to be thoroughly chloridized. For this reason an auxiliary fire is placed at the upper end of the furnace, the heat from which comes in contact with this fine dust, and passes with it under an arch, and through the flue to the dust chambers. In this way the dust is as well roasted as the ore in the ore-pit, and generally shows even a higher chlorination.

The Howell improved White furnace of 40 inches in diameter at the lower end, and 30 inches at the upper, and 24 feet long, is sufficient for any ten-stamp mill, although with some ores, also for 15 stamps. For twenty-stamp mill, the furnace must be proportionately larger. To increase the working capacity of these furnaces, it is only necessary to add to the diameter, as 24 feet is long enough for the largest size.

The proportion of the ore deposited in the dust chambers compared with that in the ore pit, ranges from 30 to 50 per cent. This division of the ore in two different places should not be considered a very great inconvenience if only a high chlorination is obtained on both sides. It takes from one to two-tenths of a cord of wood to roast one ton of ore, and from three to five per cent. of salt, which, how-ever, must be increased with rich ore, as is the case with all furnaces. The cylinder revolves from five to six times per minute, sometimes less; it depends on the quality of ore. For plumbiferous ores, five revolutions per minute would probably be too much.

Treating plumbiferous ores, the cleaning of the furnace once a week should not be neglected, otherwise the coating would increase, and finally stop the roasting. These furnaces are already introduced in a great number of mills in several States and in Mexico, being in successful operation, the chlorination of the silver ranging between 80 and 96 per cent.

O'Harra's Mechanical Furnace. This furnace was first tried in 1862 or 1863, in Dayton, Nev.; and later, three of them were built in Flint, Idaho Ty. The main feature of this furnace is the endless chain to which two oval rings are attached, the rings being as wide as the cross section of the hearth. To these rings are fastened the ploughs or shoes by which the ore is gradually pushed forward. The construction of O'Harra's furnace that was built in Flint, is shown by an outline drawing, as represented in Fig. 11. The hearth, *A*, is 104 feet long and nearly five feet wide. Eighty feet of this hearth are crossed by an arch, *B*, twelve inches high, and connected with three fire-places—two, *c* and *d*, on one side, and one between *c* and *d* on the other. *a* is the feeding hearth, provided with ore continuously from the batteries. The motion of the ore is effected by an end-less chain, *g*, passing over two chain wheels, one at each end. To this chain two oblong flat rings, *h*, are attached, each provided with eight shovels or plows, so arranged that while one of the rings shoves

FIG. 11.

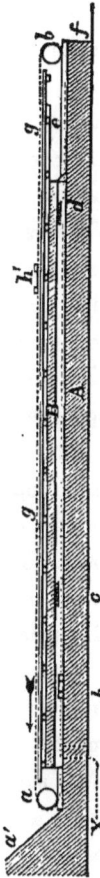

the ore toward the center line, the other pushes it back again toward the sides every three or four minutes, (or in shorter intervals if more ore is charged) The ore not only changes its place to the right and left, but it also moves forward by degrees, so that in the course of six hours from the beginning, it commences to be discharged at *f*, passing eighteen feet over the cooling hearth, *e.* On both ends of the furnace are iron doors hung on hinges, which are opened by the rings. After several months of operation the hearth or bottom appeared in good condition.

The five batteries, five stamps each, have on both long sides endless screws, by which the crushed ore is forwarded, in proportion as it is discharged, to an elevating apparatus. Being lifted about fifteen feet, it is conveyed again by endless screws along the feeding hearths of all three furnaces, *a'*, and regularly divided and discharged on the feeding hearth, *a.* The ore, mixed with 100 pounds of salt to each ton, is spread on iron plates before the batteries, (heated by the hot air from the furnaces, conveyed through the flue and under the plates.) When charged into the battery, the ore is not further handled till it comes out of the furnaces perfectly roasted.

There is only one obstacle connected with this and other mechanical furnaces. The shoes or shovels, touching the sides of the furnace, wear off by degrees, leaving a space which is taken up by the ore. This part of the ore along the wall, hardens and increases in amount in the furnace till new shoes are put in. By these the crust of one-half to three-quarters of an inch thick is broken off and carried out. From the Rising Star ore these crusts contain nearly just as much

6

chloride of silver as the well roasted ore; they are, nevertheless, disagreeable, but some means might be devised by which this inconvenience could be avoided.

The ore from the Rising Star mine, at Flint, contained argentiferous fahl ore, miargyrite, ruby silver, zincblende, galena, iron pyrites and sulphide of antimony. On an average the ore paid between $90 and $100 per ton, containing some gold. The gangue is quartz. It was crushed through sieves with forty holes to the inch, together with 5 per cent. of salt. The furnace, Figure 11, is charged continually by machinery at one end, a. The ore was moved by degrees forward, and arriving at the first fire-place, c, commenced to discharge sulphur. Between this fire-place and the second, which is on the opposite side, between c and d, the chlorination begins at an increased heat. The flame shows partly a blue color, originating from chloride of copper, and white fumes are also evolved. Between the second and the third fire-place, d, the chlorination is finished at a light red heat. From the cooling hearth, e, the roasted ore is continually discharged on the dump, f. It takes six hours before the ore from the feeding place, a, arrives at the dump. Although not more than 5 per cent. of salt is added, the roasted ore contains about 90 per cent. of the silver converted into a chloride. The gases, containing free chlorine and chloride combinations emitting chlorine, being in contact with the surface of the ore while passing over it for a space of eighty feet, have a chloridizing influence on it, replacing thus a certain amount of salt.

These three furnaces roasted twenty tons of ore in twenty-four hours. The expenses were as follows:

For wood, five cords, at $5	$ 25 00
For four men at the furnaces, at $4	16 00
For two men bringing in wood, etc.. at $4	8 00
For one man as watch in the night	4 00
For blacksmith work	5 00
For 2,000 pounds of salt, at 8c	160 00
Total expense	$218 00

or $10.90 per ton. The capacity of the three furnaces is calculated for more than twenty tons. Each one could easily treat ten tons of the Rising Star ore in twenty-four hours. The roasted ore was treated by amalgamation in pans, applying the "leaching process."

O'Harra's furnace is now greatly improved. It is built in two stories, so that when the chain comes out of the lower hearth it turns into that of the upper story. The chain is heavy, and there is no trouble whatever, either with the chain or any other part of the arrangement. Although the chain in its course through the red hot furnace is exposed to red heat, it nevertheless does not become so hot as to suffer any injury from it; there is a wooden frame work on each side of the furnace over which the chain and plows move in the open air, which prevents them from getting too hot. The furnace, of the latest construction, is eight feet wide and from forty to a hundred feet long, with four fires, two on each side, directly opposite each other. The first two fires, where the ore comes in contact therewith, are divided, so that one-half of the flame goes direct to the lower, and the other half to the upper hearth, through an opening in the arch of the fire chamber. There is a fire-clay damper to regulate the flame. The other two fires are opposite each other, so that the heat is uniform over the whole hearth. These last two fires are regulated by the ash-pit dampers.

The endless chain has two triangular frames, with plow shoes on each side. The cooling space is built in proportion to the length of the hearth. A furnace of this kind, fifty feet long and eight feet wide, can roast from thirty to forty tons in twenty-four hours with the help of only two men, consuming about two cords and a half of wood. The chlorination runs up to 90 and 95 per cent. One of O'Harra's furnaces works forty tons per day of the old Ophir tailings (Nevada), which, it is said, assay only $16.50 per ton.

The working of this furnace is not expensive, as one man can attend the roasting of forty tons in day time, and one at night.

A remarkable feature of O'Harra's furnace is the very small amount of dust that is carried off by the draught; in fact there is no other roasting furnace that could be compared with O'Harra's in this respect. Another peculiarity is the adaptability of this furnace for drying ore in pieces the size of a man's fist. One furnace dries 40 tons of ore in 24 hours, (near Shasta, Cal.) at a small expense.

The Stetefeldt Roasting Furnace. The mechanical part of the roasting itself in this furnace, is the simplest of all, and also the shortest. The finely pulverized ore, mixed with salt, is sifted continuously by a mechanical arrangement into a shaft. This shaft is about twenty-five feet high, and heated by two fire-places provided with grates. The ore, falling through the heated shaft, undergoes chlorination,—a process requiring only a few seconds. After the roasted ore has accumulated on the bottom of the shaft to the amount of about 1,000 pounds it may be drawn out. The amount of salt needed for chlorination, varies according to the ore; generally about 6 per cent., or 120 pounds to the ton, is taken, or even less, especially in treating poor ores, when half of that amount may be sufficient in most cases. A furnace having a capacity of from fifteen to twenty tons in twenty-four hours, consumes from two to three cords of wood. In twenty-four hours there are employed: Two men attending the feeding and conveying machinery, three firemen, and three men to draw and cool the roasted ore. As the latter three have time enough to carry the ore to the pans, only half of their time should be charged to the roasting expenses. According to these figures, the total expense of roasting in Reno, for instance, was not more than—

```
For labor of 6½ men at $3..........................$19 50
For Wood, 2½ cords at $6..........................  15 00
Salt, 1,800 pounds at 1½c..........................  27 00
                                                   _____
      Total expense on 15 tons......................$61 50
```

or $4.10 per ton. From 88 to 97 per cent. of the silver contained in the ore is converted into a chloride. Of the dust in the dust-chambers, the silver was found in the state of a chloride up to 96 per cent.

It is evident that with an improper treatment of the fire, by using too much or too little fuel, a less favorable result would be obtained. In the first place, if the temperature is kept too high, a part of the chloride of silver is reduced to a metallic state, which, for the purpose of amalgamation, is not so very injurious; but the metallic silver is a total loss with the lixiviation process. On the other hand, if there is not sufficient heat, some of the sulphurets may remain undecomposed. In either case the responsibility is with those in charge of the furnace; but there is nothing easier than to keep up a proper and uniform heat in Stetefeldt's furnace, there being no other hand-work about it, and all the attention of the fireman being directed to this single point.

Some ores require more heat than others; this has to be found out experimentally. Some ore gives, at a low heat, a high percentage of chlorination, but the bullion result may be very base; giving, then, a better heat, the bullion will improve a great deal in fineness without losing on the percentage of chlorination. It is essential to have the control of a good draught, and in case there should be a defect in this respect, the furnace must be examined, especially the flue, whether an accumulation of ore dust does not prevent the draught somewhere.

Generally, the bullion of the roastings in Stetefeldt's furnace is finer than of the same ore when roasted in a reverberatory, in the usual way, because, to obtain a fine bullion from a reverberatory, it requires great attention as well as much longer time for roasting, in order that all base metals be transformed into oxides; while in Stetefeldt's furnace, a larger proportion of the chloride of copper volatilizes, and all lead becomes a sulphate, according to an analysis of the Ontario ore, made by Stetefeldt.

A furnace of this kind was built in Austin, Nevada. The furnace has three important departments. 1st. The roasting shaft, twenty-five feet high and five feet wide at the bottom, narrowing somewhat toward the top, to prevent the adherence of dust to the wall. It is a simple shaft of common bricks, built as smooth as possible. On the top of the shaft is placed an iron feeder, through which a permanent and uniform feeding of the pulverized ore, already mixed with salt, is effected. The ore falls on the bottom, and when half a ton or a ton is accumulated, it is drawn out through the door. 2d. The fire-places. There are three gas generators, constructed similarly to that of the copper-refining furnace at Mansfield, Prussia. The cover is taken off and the charcoal introduced. The cover is placed again on its frame, which contains sand in a groove, in order to shut off the draught entirely. The slide door is drawn out, and the charcoal falls on the grate, through which as much air is admitted as is necessary. There are also two canals on each side of the grate. Through these canals is regulated the admission of the air for oxidizing or burning the carbonic oxide, created above the grate. In the flue, air and gas meet together, and the burning product heats the furnace. Two of these generators heat the shaft. The flue, as well as the generators above the grates, are lined with fire-bricks. The gas generators have been changed into chambers for wood fire. 3d. The dust chambers. With the draught, the gases from the shaft, with a part of the fine ore dust, pass through the vertical flue, then through the horizontal one, into a series of chambers of different sizes. The first four chambers are smaller than the four following; from the last chamber the gases draw into the chimney. The dust can be removed from the bottom of the chambers through doors. Almost all the dust is regained, and not in a raw condition, as from dust-chambers of reverberatory furnaces, requiring re-roasting, but perfectly chloridized, which is principally due to the auxiliary generator and the longer contact with the chlorine gases.

The furnace at Austin is calculated to roast from twenty-five to thirty tons of ore in twenty-four hours, at a cost of from $5 to $6 per ton, while the expenses in usual reverberatory roasting furnaces at Austin amount to $12 or $14 per ton.

It may be remarked that wherever the use of the leaching process appears admissible on silver ores, this, in connection with Stetefeldt's roasting, will allow the most economical extraction of silver, even from very rebellious ores. The baking of the ore during the chlorination, in the presence of lead, connot take place in Stetefeldt's furnace, and it is therefore very probable that a higher amount of lead will be less injurious than in any other roasting process.

Considering the old, or rather the usual, theory deduced from the roasting process in common reverberatory furnaces, that sulphates must be formed before the salt can be decomposed, and not till then will the chlorination begin, it would seem that for these chemical reactions more time is required than a few seconds; but this is not the case. As soon as ore and salt enter the furnace, each sulphuret particle ignites in the glowing atmosphere, evolving at the same time sulphur, which, in presence of the oxygen of the atmospheric air, coming undecomposed through the grates, is turned into sulphurous acid and the metal into an oxide, or in part directly into a chloride. The sulphurous acid, in contact with the ore particles and oxygen, becomes sulphuric acid. The temperature is from the start too high to permit the formation of sulphates, so that the sulphuric acid turns its force on the red hot salt particles, setting the chlorine free. All these reactions are performed instantaneously. Steam, emanating from the fuel, is also amongst the gases, consequently the creation of hydrochloric acid must ensue. The whole space in the furnace is filled with glowing gases of chlorine, hydrochloric acid, sulphurous and sulphuric acid, oxygen, steam, volatile base metal chlorides, etc.—all of them acting, decomposing and composing, on the sulphurets with great vigor. The chlorine decomposes the sulphurets directly, forming chloride of metals and chloride of sulphur; it attacks decomposingly also oxides and sulphates, if present. The hydrochloric acid performs the same office. Also metallic silver, if it should occur in the ore, would combine with the chlorine. The sulphuric acid, besides decomposing the salt, oxidizes partly the sulphurets, directly, etc.

Considering now an ore particle in a red hot condition attacked simultaneously by all these gases while falling, the final chloridizing result is inevitable. The finer the ore particles are, the more perfect the chlorination; but even if some coarser parts (to a certain degree) should reach the bottom not thoroughly chloridized, this would be finished in the pile, as the chlorination and evolution of chlorine gas continues in the red hot accumulation on the bottom of the furnace.

Since 1869 there have been erected twenty-two Stetefeldt furnaces in the United States, although a great many of these are idle now, the mines having failed to supply ore.

The latest built furnaces are greatly improved. Fig. 1 shows the construction of a furnace built at the Ontario Mill, Utah.

a is the shaft into which the pulverized ore is showered by the feeding machine, placed on the top of the cast-iron frame, b. The shaft is heated by two fireplaces. The ashpits of these are closed by iron doors, having an opening, e, provided with a slide, so that more or less air can be admitted below the grate, and, consequently, more or less heat generated. In order to obtain a perfect combustion of the gases, leaving the firebox through the slit, t, an airslit, u, connected with the airchannel, f, is arranged above the arch of the firebox. This slit also supplies the air necessary for the oxidation of the sulphur and the base metals. Another advantage of this construction is that the arches above the firebox and firebridge are cooled and pevented from burning out. The roasted ore accumulates in the hopper, k, and is discharged into an iron car by pulling the damper, l,

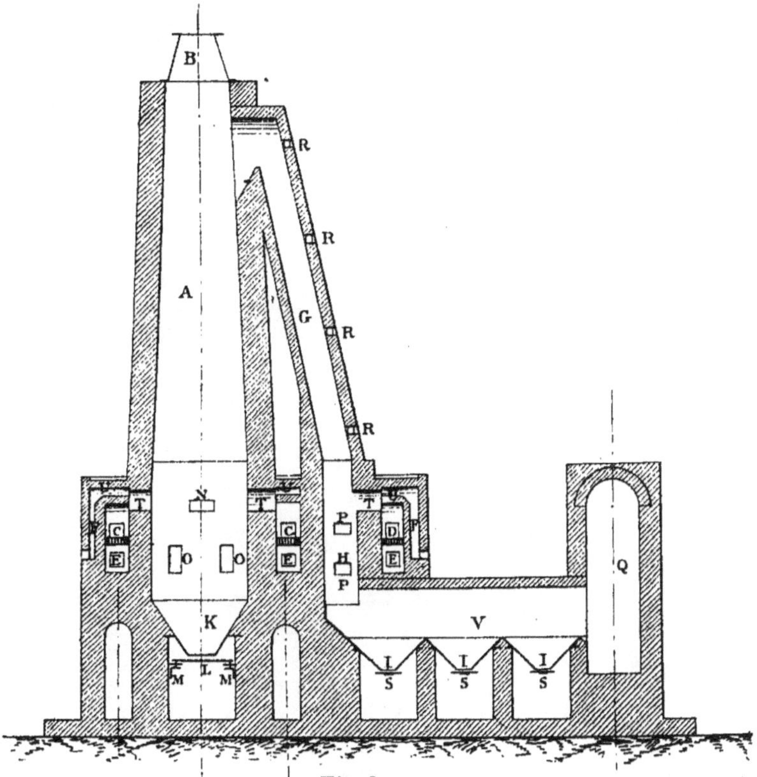

Fig. 1.

Fig. 2.

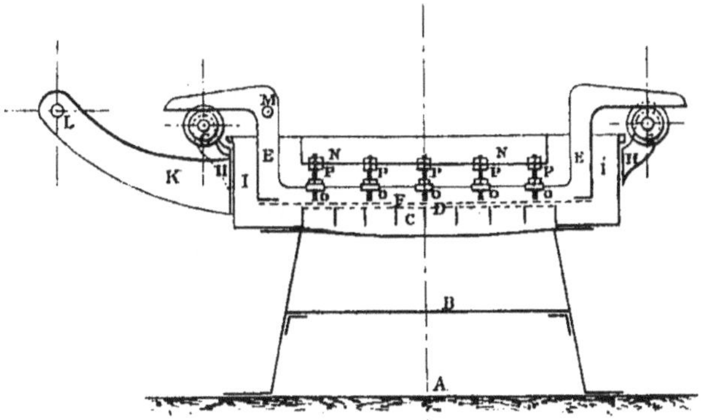

which rests on brackets with friction rollers, *m.* *n* is an observation door, and also serves for cleaning the firebridges. *o* are doors to admit tools in case the roasted ore is sticky and adheres to the walls. The gases and fine ore dust, which forms a considerable portion of the charge, leave the shaft through the flue, *g.* The doors, *r,* are provided to clean this flue, which is necessary, with some ores, about once a month. *d* is an auxiliary fireplace, constructed in the same manner as the fireplaces on the shaft, which is provided to roast the ore dust, escaping through the flue, *g,* in passing through the chamber, *h.* *p* are doors for observation and cleaning. The larger portion of the roasted dust settles in the chamber, *v,* provided with discharge hoppers, *i,* from which the charge is drawn into iron cars by moving the dampers, *s.* The rest of the dust is collected in a system of dust chambers, *q,* connected with a chimney which should rise from forty to fifty feet above the top of the shaft. At the end of the dust chambers is a damper by which the draught of the furnace can be regulated. The dry kiln can also be used as a dust chamber, and the waste heat of the furnace utilized for drying the ore before crushing it. The firing of the furnace is done on one side, and all discharges are located on the opposite side.

The Feeding Machine is shown in Figure 2. The cast-iron frame, *a,* which is placed on top of the shaft, is provided with a damper, *b,* which is drawn out when the furnace is in operation, but inserted when the feeding machine stops for any length of time, or if screens have to be replaced. *c* is a cast-iron grate, to the top of which is fastened the punched screen, *d.* The latter is made of Russia sheet-iron, or of cast-steel plate, with holes of one-eighth to one-tenth of an inch in diameter. Above the punched screen is placed a frame, *e,* to the bottom of which is fastened a coarse wire screen, *f,* generally No. 3, made of extra heavy iron wire. The frame, *e,* rests upon friction rollers, *g.* The brackets, *h,* which hold the friction rollers, can be raised or lowered by set screws, so that the wire screen, *f,* can be brought more or less close to the punched screen, *d*. The brackets, *k*, carry an eccentric shaft, *l*, connected with the shaft, *m*, from which the frame, *e*, receives an oscillating motion. To the brackets, *n*, are fastened transverse stationary blades, *o*, which come nearly in contact with the wire screen, *f*, and can be raised or lowered by the nuts, *p*. These blades keep the pulp in place when the frame, *e*, is in motion, and also act as distributors of the pulp over the whole surface of the screen. The hopper, *i*, receives the ore from an elevator which draws its supply from a hopper into which the pulverized ore is discharged from the crushing machinery. The ore is generally pulverized through a No. 40 screen. By means of a set of cone pulleys the speed of the frame, *e*, can be changed from twenty to sixty strokes per minute, whereby the amount of ore fed into the furnace is regulated. This can also be done, to some extent, by changing the distances between the punched screen, *d*, the wire screen, *f*, and the blades, *o*.

The arrangement of the feeding and conveying machinery has been lately much improved and simplified, so that no heavy and large building is required on top of the furnace, and the fireman can easily regulate the supply of ore to the feeding machine, and keep the same in running order.

Considering the enormous capacity of the Stetefeldt furnace, and its durability, if well constructed, it is by far the cheapest furnace in regard to original cost of erection.

The largest sized furnace, as represented in the drawing—the scale of which is 1 in. = 12 ft.—capable of roasting from 50 to 70 tons of ordinary ores, and from 30 to 35 tons of very base sulphuret ores in 24 hours, requires the following amount of materials, from which the cost of construction can be easily calculated by any architect or millwright, viz :

1,500 bricks, for fire boxes and arches exposed to flame.

200,000 common bricks, of good quality, for furnace, large system of
dust chambers, chimney, and cooling floor.

2,500 ℔s. in bolts and nuts for anchoring furnace and dust chambers.

4,500 ℔s. in wrought iron braces, flat iron for car-guides, tools, etc.

16,000 ℔s. in castings.

All the castings are very plain and simple, the water-jacket on top
of furnace, and the water damper, having been discarded. Considerable work is only required on the feeding-machine, feeding-machine
damper, and discharge damper, and some on the fire-doors, which will
be covered by an additional charge of about $700 added to the ordi-
nary price of castings.

The cost of three iron discharge cars is $125. For furnace of 15
to 20 tons capacity, without hopper discharge, and a less extensive
system of dust chambers, the amount of materials required may be
estimated at two-thirds of the figures given above.

The cost to erect a Stetefeldt Furnace of largest size, at stated prices
for materials, freight and labor, is the following:

1,500 fire-bricks, at $60 per M	$ 90 00
200,000 common bricks, at $15 per M., delivered...........	3,000 00
7,000 ℔s. wrought iron in bolts, braces, etc., at 8 cts.......	560 00
16,000 ℔s. castings, at 6 cts	960 00
Labor on castings.......................................	700 00
Iron discharge cars.....................................	125 00
Freight on 12,000 ℔s. fire-bricks, 23,000 ℔s. iron, at 2 cts..	700 00
225 days' mason labor, at $6.00........................	1,350 00
180 days' mason's helper labor, at $3.00	540 00
To grading, foundation, sand, centres, and scaffolding......	600 00
Superintendence of construction........................	400 00
Total..	$9,025 00

This estimate does not include cost of conveyors and elevators, and
the building to cover the furnace. A part of this expense belongs, in
reality, to the pulverizing machinery. These items will vary materi-

ally according to the price of lumber. From \$3,000 to \$4,000 is a liberal estimate for most mining districts.

The expense of roasting in the Stetefeldt Furnace can be calculated as follows:

FUEL.—The amount of fuel required in 24 h. is from 2 to 4 cords of good wood, (or its equivalent in coal) according to capacity of furnace and character of ore. Furnaces of 20 to 25 tons capacity, have generally consumed from $2\frac{1}{2}$ to $2\frac{3}{4}$ cords of nut pine. Only well seasoned wood should be used for firing the furnace.

LABOR.—Two firemen take charge of the furnace in 24 h., in 12 h. shifts. For discharging and cooling the roasted ore, the number of men required is in proportion to the amount of ore roasted. For instance, a furnace of 25 tons capacity requires four men in 24 h. to discharge and cool the pulp, and get it ready for amalgamation.

SALT.—For chloridizing roasting of silver ores, the Stetefeldt Furnace requires less salt than any other furnace, because the decomposition of the salt is very perfect, and the chlorine and chloridizing gases, emanating from the roasted ore at the bottom of the shaft, act upon the falling ore, which floats in an atmosphere of these gases. Ores which are free from lime and magnesia can be chloridized in this furnace with a minimum of salt. The results of a run of nine months of the Surprise Valley Mill, Panamint, Cal., where an average chlorination of 92 to 93 per cent. was obtained in roasting silver ores, of \$75 assay value per ton, with only $2\frac{1}{2}$ to 3 per cent. of salt. In fact, a still lower percentage of salt would have been sufficient, had it been an object to save more. In case the ore carries lime and magnesia, or a larger percentage of base sulphurets, it is, of course, impossible, even in the Stetefeldt Furnace, to obtain good chlorinations with such a slight percentage of salt, and generally from 5 to 8 per cent. are required. The salt is either mixed with the ore on the dry-kiln and both crushed together, or ore and salt are pulverized each separately, and mixed by proper machinery before entering the Stetefeldt feeder.

The cost of roasting in a Stetefeldt Furnace of 25 tons capacity, in 24 h., at stated prices for labor, fuel, and salt, such as are generally paid in mining districts of Nevada, are calculated as follows:

```
2 fireman, at $4.50........................................$ 9 00
4 pulp-coolers, at $4.00 ..................................  16 00
2¾ cords of wood, at $8.00.................................  22 00
Wear and tear of screens, etc..............................   1 00
                                                            ———
   Labor and fuel for 25 tons............................$48 00

Labor and fuel, per ton..............................$1 92
7 per cent. salt, at $40.00 per ton.....................  2 80
                                                        ———
   Expense of chloridizing roasting...............$4 72 per ton.
```

The percentage of silver chloridized in roasting silver ores in the Stetefeldt Furnace varies according to the character of ore, and the care with which the furnace is managed. Results as high as 97 per cent. have been obtained, while the average chloridations generally range from 87 to 93 per cent. Ores free from sulphur, or with only a slight percentage of it, should be mixed with one or two per cent. of pyrites of iron, otherwise decomposition of the salt is not possible. Oxidized ores, however, carrying peroxides of manganese and iron, which give off oxygen, can be successfully chloridized by themselves. The best results are obtained by mixing sulphuret ores with oxidized ores, mainly if the latter contain peroxide of manganese, whereby also the capacity of the furnace is much increased.

The presence of copper is very favorable for chlorination of the silver, and if the ore is of such a character that it bears a high heat in roasting without sintering, the chloride of copper, formed in the upper part of the shaft, can be almost completely decomposed, and very fine bullion produced by amalgamation. As an example, it may be stated that the results of an experiment made at the Surprise Valley Mill, Panamint, Cal. The ore, roasted at a low temperature, gave bullion only 300 to 400 fine by amalgamation, the base metal being copper.   By roasting the same ore at a high temperature, the bullion produced was almost free from copper, its average fineness being 980 during a run of nine months.

All antimonial ores are chloridized with great facility, and with a good system of dust-chambers the loss of silver by roasting is hardly perceptible.   The same is the case with zincblende.

A Stetefeldt furnace was put in operation at the Ontario Mill in February, 1877.   The Ontario mine carried in its upper levels a great amount of decomposed ore, which yielded a very high percentage of its silver by raw amalgamation.   In the lower levels, however, the character of the ore changes entirely, and the ore from this part of the mine can only be successfully amalgamated by first roasting it with salt.   After reaching the 500-foot level the ore became very much baser than that met with before, and as the working of the furnace gave occasionally much inferior results than those formerly obtained, an analytical investigation was ordered.

Specimens from the 500-foot level of the Ontario plainly show the following minerals, viz: zincblende, galena, fahlore, pyrites of iron. Of the zincblende there are two varieties, the one of a light yellow color, the other black.   In the latter a larger percentage of fahlore is found between the cleavages than in the former.   Also, the galena is often intimately mixed with fahlore.   The fahlore itself occurs, besides, in compact masses.

A chemical analysis of an average sample of crushed ore, made by Stetefeldt, gave the following results:

One hundred parts of the ore contain:

    9.60 zinc,
    6.07 lead,
    2.77 iron,
    1.41 copper,

0.45 manganese,

0.60 silver,

7.68 sulphur,

1.20 antimony,

0.20 arsenic,

55.21 silica,

13.14 alumina,

1.00 potassa and soda from decomposed feldspathic matter.

Traces of bismuth, cadmium, lime, magnesia.

Assay value, $229.34 silver per ton.

The silver-bearing minerals, after eliminating the fahlore from the zincblende and galena, were found to be composed as follows :

Yellow zincblende contains:—1.60 per cent. iron,

0.40   "   copper,

Trace cadmium.

Assay value, $92.50 silver per ton.

Black zincblende contains:——2.40 per cent. iron,

0.84   "   copper,

Trace cadmium.

Assay value, $92.50 per ton.

Galena contains..............1.80 per cent. iron,

0.66   "   copper,

Trace bismuth.

Assay value, $175.00 silver per ton.

The fahlore contains.........24.16 per cent. copper,

11.77   "   silver,

6.70   "   zinc,

4.20   "   lead,

27.00   "   antimony,

4.40   "   arsenic,

22.20   "   sulphur.

Assay value, $4,440.00 silver per ton.

The fahlore is the main source of the silver in the Ontario ore. It
is, of course, more or less intimately mixed with the galena and zincblende, but in crushing the ore through a No. 60 screen, the fahlore
becomes sufficiently free to allow its roasting independently of the
minerals with which it is associated.   Ore of exactly the same character as analyzed was roasted with about 13 per cent. of pure salt in
the Stetefeldt furnace, and samples taken from the shaft and flue.
These were subjected to a chemical analysis.   The object of this investigation was to ascertain if there is a sufficient quantity of salt
falling down the shaft, as the chlorinations there have been recently
often very low ; then whether undecomposed salt is left in the roasted
ore, and how much, also which of the base metals have been changed
to chlorides and which to sulphates ; finally, how much sulphur, not
changed to sulphuric acid, has been left in the roasted ore in combination with metals as sulphurets.

The solution of these questions must give an exact knowledge of
the working of the furnace.

The sample of roasted ore from the shaft was found to contain in
100 parts:

    0.25 chloride of copper,
    1.51 chloride of aluminium,
    1.38 chloride of zinc,
    3.68 chloride of sodium,
    Traces of chlorides of other metals,
    3.26 sulphate of lead,
    0.56 sulphate of alumina,
    4.62 sulphate of soda,
    Traces of sulphates of other metals,
    Rest, metallic oxides and gangue.

Of the silver contained in the ore, 81.32 per cent. were chloridized.
Sulphur in undecomposed sulphurets, 0.18 per cent.

The sample of roasted ore from the flue was found to contain in one hundred parts:

1.07 chloride of aluminium,
3.08 chloride of sodium,
Traces of chlorides of other metals,
0.74 sulphate of copper,
2.88 sulphate of alumina,
1.48 sulphate of zinc,
5.18 sulphate of lead,
10.01 sulphate of soda,
Traces of sulphates of other metals,
Rest, metallic oxides and gangue.

Of the silver contained in the ore 82.24 per cent. were chloridized. Sulphur in undecomposed sulphurets, 0.064 per cent.

These results are very interesting, and prove:—

1. That in the shaft, as well as in the flue, over three per cent. of undecomposed salt are present, and that, consequently, an increase of salt in roasting would be useless.

2. That there remains more undecomposed salt in the shaft than in the flue, and that the low results in the chlorination of the silver frequently obtained of late in the shaft cannot be accounted for by an insufficient quantity of salt dropping down the shaft. Consequently, it would be of no benefit to crush the salt by itself through a coarser screen.

3. That the volatilization of salt is by no means as considerable as supposed.

4. A peculiar relation exists between the shaft and flue in regard to the salts of copper, zinc, and aluminium. Copper and zinc are present in the shaft only as chlorides, in the flue only as sulphates. The aluminium exists in the shaft as well as in the flue in both combina-

7

tions, but in the shaft the chloride is predominant, and in the flue the sulphate. It is evident that in the shaft the metallic sulphates are more completely decomposed, as the ore, falling against the rising gases, is longer suspended in the heat, and passes finally through the hottest zone of the furnace. In the flue the ore particles move with the gases, are only for a moment exposed to the flame of the auxiliary fireplace, and, being instantaneously transferred to a lower temperature, the sulphates, once formed, have no occasion to lose their acid, except by contact in the charge with salt, which is, necessarily, a slow and imperfect mode of decomposition. The presence of these sulphates has, however, a beneficial effect upon a further chlorination of the silver in the ore at the bottom of the flue.

As to the lead, we find it only as a sulphate in the shaft as well as in the flue, but the preponderance of the formation of sulphates in the flue becomes again evident. As the sulphate of lead does not lose its acid at a high temperature, it follows that its decomposition in the shaft cannot be accomplished.

5. The complete oxidation of the sulphur is very surprising, and shows how thoroughly the furnace does its work even with such base ores.

6. Chlorinations of eighty-one and eight-two per cent. of the silver are very high for Ontario's ore, and with such chlorinations from ninety to ninety-four per cent. of the silver ore is extracted by amalgamation without any difficulty.

The results of this investigation have clearly demonstrated that ores like those from the 500-foot level of the Ontario mine, can be worked with perfect success by the Stetefeldt furnace, if proper conditions are maintained. Inferior results were only caused by a deficiency in the draught of the furnace, or in overfeeding it when the draught was not in a normal condition. These defects can be easily corrected.

III. EXTRACTION OF SILVER

BY WAY OF LIXIVIATION.

To lixiviate, or to leach a soluble metal out of ore—that is, to filter a liquid through the ore, so that it dissolves and carries out the metal in a clear solution—is the process called the lixiviation, or leaching process. If there is, for instance, sulphate of copper, and sulphate or chloride of iron in the ore, the water, passing through it, dissolves the soluble copper and iron salts, and takes it out of the ore; but it does not dissolve the chloride of silver; therefore this remains with the ore in the leaching tub. In order to leach now, also, the silver, another liquid must be employed, which has the property of dissolving the chloride of silver. Chloride of silver is the only silver combination that can be successfully lixiviated. It is therefore a primary condition that the ore should be subjected to a chloridizing roasting, before the leaching process can be made use of. No silver ore, except natural chlorides or bromides, is suitable for lixiviation in a raw state.

There are different solutions which dissolve the chloride of silver. A hot concentrated solution of common salt, a solution of hyposulphite of soda or of lime, ammonia, etc.; all dissolve the chloride, and filter with it as a clear solution, out of which it must be precipitated. There are again different ingredients by which this can be effected. The silver is precipitated by polysulphide of sodium, or of calcium, also by sulphureted hydrogen, which is cheaply produced by melting in a porcelain or glass vessel parafin, with flower of sulphur. In all these cases the silver is precipitated as a sulphide, or in metallic condition by metallic copper,

The extraction of the chloride of silver by alkaline hyposulphites was proposed by Percy. Patera was the first who made use of the hyposulphite of soda for extraction of silver in a practical way; his success, however, depends principally on his modified and complicated roasting. By lixiviation the silver is extracted in the Patera, Kiss, Roszner-Patera, Ziervogel and Augustin processes.

The extraction of silver by the solving process is simple. The ore is first roasted with salt in the usual way, whereby the formation of base metal chlorides cannot be avoided entirely. After roasting, the ore is first subjected to leaching with water, in order to extract the base metal chlorides, and then with hyposulphite of lime, to extract the silver.

The belief that by the leaching process a purer bullion can be obtained than by amalgamation is erroneous. If there is a great deal of lead in the ore, which in roasting changes into sulphate of lead, this will not be amalgamated in pans, while all of it, may it be as a chloride or as a sulphate, will be dissolved by the hyposulphite and carried out with the silver.

The only way under such circumstances to get a fine bullion by the leaching process, is explained further on below.

The Extraction of Silver.

After a chloridizing roasting the ore should be examined to ascertain the amount of chloride of silver contained in it (page 32.) In case the extraction should not be satisfactory, it is then easier to find what the cause is.

The leaching vat is best made round, as it is much easier to make the staves fitting, water tight, than large square boxes. Large vats can be made also of masonry, lined with asphaltum. The bottom in this case must be made very carefully. In order to permit the filtration, there must be a false bottom in each vat or box. This false bottom is prepared in different ways. On the bottom are laid pebble stones of the size of a hen's egg, then on this a layer of smaller ones, and so on till all is covered with sand free of mud; this false bottom is 4 or 5 inches deep. The ore to be leached is thrown direct on this filter. A better way is to put wooden staves of about one inch in thickness, as represented in figure 14.

Fig. 14.

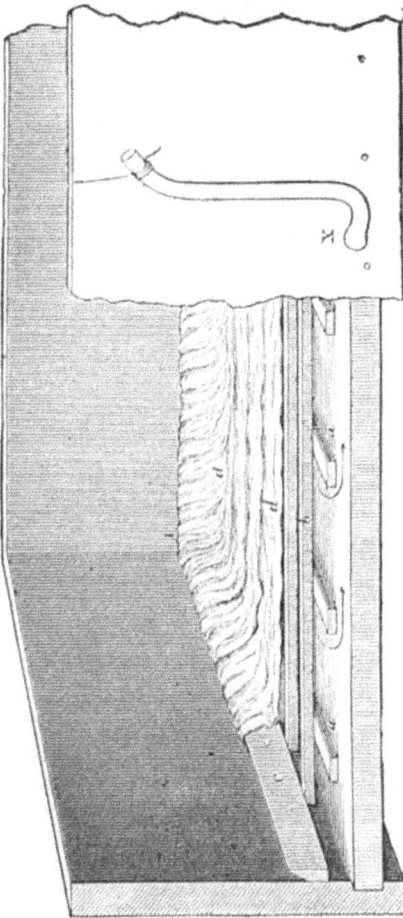

The first staves, *a*, on the bottom, are 12 inches apart, and on the front side where the outlet, *x*, is, the staves should not touch the side, so that the solution can flow to the outlet, as shown by the arrows. The second row of staves is layed across the first ones, one inch apart. The whole is then covered with a strong, but not too tight, cotton cloth, *d*, or some other stuff like gunny sack, cocoa matting, etc. It happens, that when water is admitted, this breaks through in one or the other place, and carries out ore. This can be effectually prevented by fixing above the staves, on all four sides, a slat of four inches wide, *c*, cut as represented in the drawing. The ends of the wet cloth are pressed into the space formed by the beveled cut. The size of leaching vats is generally from 8 to 10 feet in diameter; or, if rectangular tanks are preferred, they can be made as large as 12 by 15, or 12 by 12 feet. The last sized tank would then contain 144 square feet, and if the ore is charged one foot deep, there are so many cubic feet. The roasted ore generally weighs from 65 to 75 lbs. per cubic foot, and, calculating the average with 70 lbs., it would require about ten tons of roasted ore to charge a like tank; but some ore allows filtration when two feet deep, and in this case the charge would be doubled. From bricks, lined with asphaltum, such tanks can be made large enough for ten to twenty tons of ore at a charge. The hight of the sides must corres- pond with the nature of the ore. Some of it filters freely if 25 inches deep, but other kinds of ore hardly allow filtering if it is charged only 12 inches deep. There should always be at least 12 inches space above the charge to receive the solution. In the outlet, *d*, an india rubber hose of two or three feet in length is tightly inserted, and the end of it hung up, as shown in the sketch.

Generally, the vat is charged with the roasted ore without sifting, and the surface spread evenly, leaving about six inches space from the top. The water is now introduced, either hot or cold, and if the filtering goes very slow, hot water will help a great deal. It sinks slowly towards the bottom, and the air escapes through the hose. After the ore is covered several inches with water, the hose is discon- nected and placed in the trough which conveys the lixivium to the precipitating tubs.

In the beginning, the leaching water, as it escapes through the hose, is highly charged with base metal salts, and shows a green color if there is much copper in the ore. The water is kept running continuously, and the influx and the efflux are equalized. After one or two hours a glass full of the liquid, at the hose, is taken, and a few drops of sulphide of calcium (or of sodium) added. If a precipitate falls, of a dark or light color, the leaching must continue; but it is not necessary to continue until no precipitate at all is perceived, as it requires some time—perhaps an hour, before all the water runs out. The water which comes out last must be free from salts. This first leaching takes from two to four hours, sometimes longer.

As soon as the ore is freed from the base chlorides soluble in water, a solution of hyposulphite of lime or in the commencement, a solution of hyposulphite of soda is led in from a tub or tank, on the ore, in order to dissolve the chloride of silver. This leaching is conducted like the former. It depends on the amount of silver how long this work continues—from eight to twenty hours. The clear cold solution, containing the chloride of silver in the form of a double salt, has a very sweet taste, and is conveyed through a trough or india rubber hose into a precipitating tub. Very rich ore, containing from 12 to 15 per cent. of silver, would require forty-eight hours leaching, and even then it would be necessary to subject the ore to a second leaching with the hyposulphite, with an intermediate roasting with green vitriol and salt; for, with the best work, if 95 per cent. are extracted, the tailings would still appear sufficiently rich for this, containing about 200 ounces of silver per ton. Ores containing $350 per ton are often leached out perfectly in twelve hours. The end of the lixiviation is ascertained in the same way as in leaching with water, using the sulphide of calcium. If no precipitate is obtained, the extraction is finished.

The tailings, as they are now in the vat, contain a great deal of hyposulphite solution, which should not be lost. Water is therefore turned again on the tailings to displace the solution, which is conveyed into one of the precipitating tubs, as long as there is a distinct taste in it, and in the same way, when leached with water before the silver was extracted; the water must be displaced by the hyposulphite before the solution is turned into the precipitating tubs. This part, however, must be carefully attended to, not to miss the time when the silver begins to come out; a sweetish taste indicates the time, or a slight precipitation with hyposulphide of calcium.

The color of the precipitate is a black-brown. The presence of base metals changes the color somewhat. Iron makes it black; copper, red brown; lead and antimony, light red-brown, etc. The silver is first dissolved, especially if a diluted solution of hyposulphite of lime is used, and for this reason the first precipitate is the richest in silver. Ore containing a great deal of lead—especially if the roasting was so conducted that a large part of it remained as sulphate of lead, which is not soluble in the leaching water—will give in the beginning of the leaching with the solvent a precipitate of silver with some lead; afterwards, however, the silver diminishes, so that the precipitate of lead finally appears free of silver. Besides the sulphate of lead, sub-chlorides and oxy-chlorides are formed during the roasting, which are not soluble in water, but are dissolved by the hyposulphite of lime; for this reason always some base metals will be found in the precipitate.

In case rebellious ores are treated, and hot water is used for the extraction of base chlorides, a better silver is obtained if the ore is cooled down by cold water before the cold and diluted solvent is applied. Purer ores may be treated with a warm solution of the solvent.

When examining the tailings as to the amount of silver left therein, as compared with the original ore after roasting, it must be remembered that, after leaching out a quantity of metals by water and the solvent, the ore lost a considerable part of its original weight, and that, consequently, one-half ounce of such tailings taken into assay will always give a larger silver button than there ought to be. A true assay of leached tailings is made if half an ounce of the same ore is leached on a filter with hot water and hyposulphite of lime, in the same way as the ore on a large scale, washed with water, dried and weighed. The weight found after leaching must be taken for half an ounce in assaying the tailings.

The residue, or tailings in the leaching box, must be removed now as valueless. The sides of the leaching boxes are from eighteen to twenty-four inches above the bottom, and being from six to eight feet square in the clear, the removing of the tailings by means of shovels is easily effected. The men must be careful not to dig too deep, otherwise the filter will be injured. It is quite proper to fix wooden staves, as long as the box requires, on top of the filter. These staves are one inch wide and one-half of an inch thick, and are placed from four to five inches apart, so as to protect the canvas or filter against the shovel.

If the ore is very base, the greatest part of the silver can be obtained over 800 fine, if the solution is not strong ($\frac{1}{2}$° Bau.). When the leaching commences, the solution appears sweetest from dissolved silver. After a while, when the sweet taste disappears, the leaching may be continued for one-quarter or half an hour, and then the following solution conveyed into another precipitating tub. This last precipitation will give a very low bullion, which should be either cupelled with lead, or if too poor, dried and roasted again with ore. The time when the lixiviation should cease, is easily found out. With the watch in hand, the solution from the leaching vat is led in a bucket, or better into a pitcher, for exactly one minute. The silver and other metals are precipitated by sulphide of calcium, collected carefully in a filter, washed with water, and then, after drying thoroughly, melted with litharge in a crucible and cupelled. Sixty times the weight of the resulting button gives the amount of silver per hour that is leached out. A calculation for the day shows whether it covers the expenses of labor.

There is another important fact connected with the lixiviation. The different chlorides, being removed in the first leaching with water, are principally those of copper, iron, lead, antimony and zinc, besides some undecomposed salt. The first quantity of water introduced into the leaching box is, of course, most saturated with the named salts, and they have the property of dissolving, also, some chloride of silver. The dissolved silver precipitates again as soon as it becomes diluted with more water. There is, therefore, no difficulty in regaining the silver which is thus leached out. The amount of silver carried out by the leaching water varies from 0.5 to three per cent. Not only the chloride of silver, but also those of lead and antimony, are precipitated by dilution with water. There are two ways of regaining this silver.

Mr. O. Hofmann adopted an ingenious plan for this purpose, by conveying the hot water, under a slight pressure from below, under the false bottom, as described in (page 124.)

The other plan is the precipitation of the silver, together with the chlorides of lead and antimony, outside of the leaching box. This mode is preferable to the former when a great deal of lead and antimony is in the ore; for if precipitated in the box, all of it will be dissolved by the hyposulphite of lime, and then precipitated as sulphides with the silver, making this impure and consuming much of the precipitating agent. As soon as the dissolved chlorides flow into the trough, below the leaching vat, into which several leaching boxes discharge their fluids in different degrees of dilution, the gradual precipitation commences. An additional small stream of clear water will hasten the precipitation, which is white and adheres to the trough through the whole length of it. These chlorides are the richest, and contained, at Flint, Idaho, 9 per cent. of silver; the balance was principally lead and antimony. The precipitate deposits on all bodies offering a surface. For this purpose a box must be constructed, the sides of which are six inches high. There are partitions four inches high, leaving a space of six inches between them, so that the water must flow from the first partition, which on one side does not reach the six-inch side of the box, into the second, and from this on the other side into the third partition, and so on in a zigzag way. The space between the partitions is filled with shavings, offering a large amount of surface for the chlorides to deposit thereon.

The water leaving, the box contains, principally, copper and iron in solution.

The white precipitate, when accumulated, is taken out, placed in filtering bags, with or without the shavings, and washed with clear cold water, in order to get rid of the copper solution. The silver can be extracted in two ways: The simplest mode is the application of hyposulphite of lime. The sediment is taken out from the filtering bags and charged, while wet, into a filtering box of a proper size. The hyposulphite of lime, in a cold condition, is poured over it and managed like the ore. The silver-holding fluid may be conveyed into the precipitating tub and treated with the solution from the ore. The liquid from the filtering box is examined from time to time with sulphide of calcium. In the beginning the precipitate appears dark, being mostly silver; but when it is perceived that the precipitate assumes a light yellow color, too much of lead, zinc and antimony is being carried out, and the leaching must be stopped. The residue in the filter box contains still some silver.

The other mode of extraction is more perfect, but also more expensive and more troublesome. After the copper has been washed off, the contents of the bags are taken out and dried. It is then introduced into large crucibles and smelted with an addition of soda-ash. The reduced metal, if some lead occurs in the ore, must be separated by means of cupellation, resulting in clean silver and litharge.

The chloride of copper, after the silver with the lead deposited in the trough and box, is led into a reservoir in which old iron is suspended. The copper precipitates in a metallic state on the iron, and about eighty-eight parts of iron go into the solution in place of one hundred parts of copper; consequently, as each one hundred pounds of pure

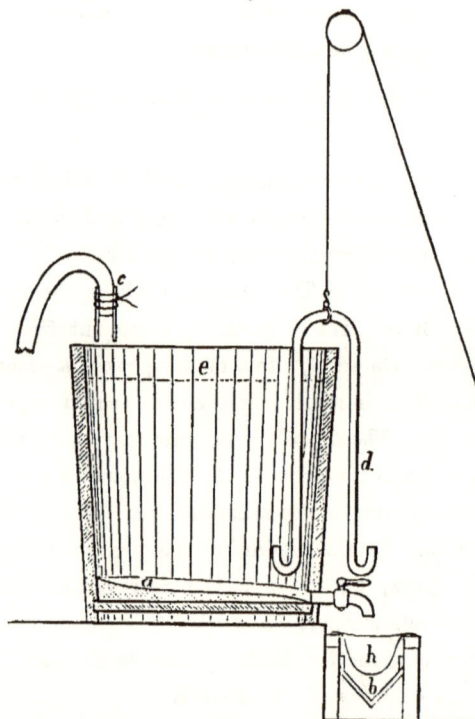

Fig. 15.

Treating always the same kind of ore, the required quantity of the precipitating agent is soon learned.   The black precipitate sinks to the bottom, and the workman now dips a little of the clear liquid out in a glass tube or tumbler, and adds a few drops of the sulphide of lime.   If a dark precipitate or a dark color is produced, it shows that there is still silver in the liquid, and more of the agent must be added; but if, on the contrary, no precipitate is observed, there is either enough or too much of the sulphide.   To prove this, some of the silver-holding liquid is added to a test, taken from the other tank under treatment.   If in this case a precipitate is formed, silver-holding liquid must be carefully added to the liquid in the tank until no reaction is produced. This work, delicate as it seems, is easily learned by the workmen. If a little silver should be left in the liquid, it is not injurious, neither is the silver to be considered as lost, because the same liquid is used over again ; but a small excess of the sulphide of calcium would cause a loss in silver, as it precipitates sulphide of silver in the leaching tank in the mass of ore, which is not dissolved again.   The precipitation is performed in a short time, requiring about fifteen minutes for each tank.   The stirring must be executed with vigor.   Wooden grates fixed to a vertical stem will answer the purpose.   Some use perforated wooden disks on the stem.   After the last portion of the precipitant had been added and well stirred, the precipitated silver will soon settle; but it is well to allow one hour's time at least.   The liquid must be clear.   All the sulphide of calcium that was used, by giving up a part of its sulphur to the silver, is reduced to hyposulphite of lime, and in this way the amount of hyposulphite is constantly increasing; but on the other hand there is also some waste occurring at the end of the leaching.   If the sulphide of calcium is made too weak, the hyposulphite of lime becomes diluted by degrees, and the lixiviation would be imperfect.

The clear solution above the settled precipitate must be drawn off, either with a rubber syphon, to one end of which two short pieces of wood are fastened, as shown in figure 15, c, so as to prevent the hose from touching the bottom, in which case it would draw precipitated silver; or the clear solution can be removed also by more convenient syphon, shown by d, which is made of a lead pipe.   By lowering the same, the clear solution commences to flow through the outside end, as soon as it sinks below the level of the solution.

The solution from the precipitating vat runs through the gutter, b, figure 15, into a large vat, from which it is pumped up into the solution vat, standing above leaching vats.   The hyposulphite of lime is thus constantly circulating from the uppermost solution vat through the leaching vat into the precipitating tubs, and from thence into the lowest receiving vat, again to be pumped up into the solution vat. For this purpose wooden pumps answer very well.

Treatment of the Precipitated Silver.

After the precipitated silver has accumulated to 10 or 12 inches deep, and the clear solution of hyposulphite of lime has been removed, the precipitate can be taken out by the hole at the bottom of the vat provided with a large brass cock, as shown in Figure 15.

Each precipitating tub has a filter of cotton cloth, *h*, tacked on a wooden frame about two feet wide and from three to four feet long, or differently shaped filters like those used for amalgam. Into these filters the precipitate is allowed to flow until almost full, and after the solution has filtered through and the precipitate settled, more of the silver is allowed to flow in, this being repeated several times till the filter remains well charged. After this, clear water is admitted till all hyposulphite of lime is removed and no taste is observed in the filtering water.

After several hours the black precipitate must be pressed to get out the remaining water. A screw press answers well for this purpose, but the precipitate should be wrapped up in a cloth and then pressed. The black silver cakes are then taken out and dried in a warm room, or in a drying oven. For the purpose of burning off the sulphur, the dried sulphide is introduced into a muffle or other calcining furnace, and heated till the sulphur commences to burn with its known blue flame. When this disappears the heating must continue at a dark red heat for several hours. By this operation the cakes are reduced almost entirely to metallic silver, generally covered with threads of silver; sometimes an intense green color is assumed by pieces remaining in the furnaces over night.

The burned cakes are now prepared for smelting in crucibles. They are placed in black lead crucibles, according to the size, up to three hundred pounds, and fused. All the sulphur was not driven out by the preceding operation. The remaining part must be removed by placing metallic iron in the fused metal; thereby iron matte is formed which rises to the surface and is skimmed off. The surface of the silver is then cleaned by adding some bone ash and borax, or borax alone, which is also skimmed off and the silver dipped out or poured out into moulds. According to the careful treatment in the roasting process, and the nature of the ore, the silver will be from 800 to 950 fine.

Mr. O. Hofmann, in want of sulphur for the production of sulphide of calcium, used to calcine the dried sulphide of silver in iron retorts. In this way he obtained a large proportion of sulphur as a fine sublimate. This could be done also in a proper muffle furnace, so arranged that, after all obtainable sulphur had sublimated in a receiver, this could be removed, and the calcination continued under access of air. At Melrose, O. Hofmann tried to dry the silver precipitate by a centrifugal machine, with excellent result, in less than one minute.

Applicability of the Lixiviation Process.

All silver ores which allow a good chlorination in the furnace, and which permit filtration, are suitable for the lixiviation process. The great advantage of this process is cheapness. Roasting, of course, is indispensable, except with chloride ores; but neither pans and the required power, nor quicksilver, are used, and for this reason less capital is necesary to put up reduction works. All the cupreous silver ores of Cerro Gordo, Yellow Pine, Montgomery, and of the other new silver districts, can be treated to great advantage by the solving process. But not only the dispensing of pans, power, and whatever belongs to them, deserve consideration. In using pans, there is gearing, belting and other parts exposed to breakage; the pans wear out in time, especially with base roasted ore. The cost of quicksilver

8

itself is not great, provided it has not to be transported too far; but this process is important in localities where lime and sulphur can be got in the neighborhood, while the transportation of quicksilver is troublesome and costly. The process, although a little more complicated than the pan amalgamation, and requiring more time, is nevertheless simple and easily learned.

Ores containing much clay and lime, or talk, filter very slowly, sometimes so slowly that pan amalgamation must be preferred. The remedy in such case is found in separation of the mud or slime from the coarser sandy part. The crushing, therefore, must be changed to wet crushing, and the separation must be effected either through pointed boxes or some other arrangement, where the flow of water is swift enough to carry the fine stuff further than the sandy part. This latter, for the purpose of roasting, must be dried. Generally, it is charged wet in a roasting furnace with two or three hearths. In this case, the salt is added when the ore is in the last hearth. But then comes the inconvenience of handling the slimes. In Mexico, this part can be subjected to the patio amalgamation, provided the silver comes out by this process, and in the United States the slime, on account of its extreme fineness, may admit a direct pan amalgamation; but in either case a double manipulation is very inconvenient. The proposal to treat such ore without separation in an agitator with a filtering bottom, may perhaps answer in some instances; but this point must be decided by experiments.

Ore that permits filtration, but too slowly, can be leached much quicker with hot solutions, to begin with hot water; but the hot water in the beginning dissolves more chloride of silver than cold water.

Lixiviation with aid of suction is a great advantage where it can be arranged. The leaching boxes must be located ten to twelve feet above the precipitating vats. The arrangement is represented by Figure 16.

Fig. 16.

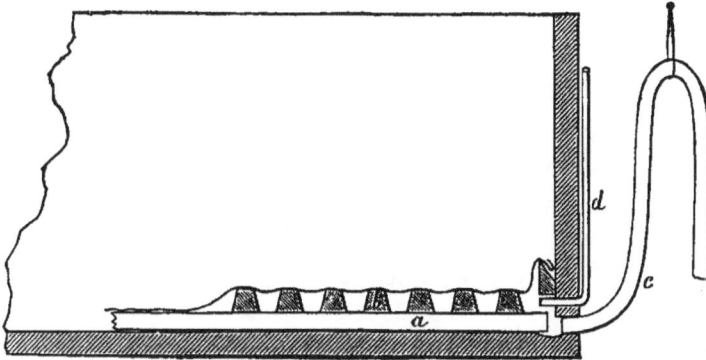

The hose, c, is inserted at the lowest point of the vat below the false bottom ; then there is a glass pipe, d, reaching outside near the rim of the vat. The hose, before commencing, is likewise lifted as high as the glass tube. a represents the first staves on the bottom of the vat, and b the second row which is covered with the cloth. After the vat is filled with ore and the water admitted, the air below the false bottom escapes through both pipes ; but after a while, when the water rises and covers the mouth of the hose, the air from between the upper staves goes through the glass tube and is followed by the water. When this reaches the water level inside the tank, the tube is shut up by a tight stopper and the hose lowered down to the precipitating vats. The lixiviation is much quicker than without suction, but if the hose should be so wide that more water escapes than can follow, there will be, of course, no suction.

Hyposulphite of Soda.

The hyposulphite of soda is a cheap commercial article of white transparent crystals (5 to 6 cts. per ℔.). It is made in different ways. Four parts of glauber salt (calcined) are mixed with 1 to $1\frac{1}{4}$ parts of charcoal or coal dust moistened a little, and introduced into a large crucible or in an iron vessel, and then exposed to a red heat from six to ten hours. After this, it is pulverized, moistened again, brought in contact with sulphurous acid in thin layers, lixiviated with water, and the water evaporated sufficiently to make the salt crystalize. A cheaper method is to use, for this purpose, lime which has been used in gas factories for purifying gas. This lime, if exposed to the action of the air for some time, is changed into hyposulphite of lime, and is easily decomposed by carbonate of soda. But the cheapest way for metallurgical purposes is to buy this article, as several hundred pounds is all that is needed to commence with. Afterwards, if carefully handled, by the use of sulphide of calcium, the quantity of the hyposulphite increases, and as there is always some waste of soda replaced by lime, the whole liquid is changed by degrees into hyposulphite of lime. If there is a small amount of gold in the ore, the hyposulphite of soda does not dissolve the gold, which, after roasting, is converted into a combination with chloride of sodium; but the hyposulphite of lime does dissolve it. In this case, if there is gold in the ore, it may be of interest to commence directly with the lime solution. The

Hyposulphite of Lime

is best prepared by taking three parts of lime to one and a half parts of flour of sulphur and boiling it either in an iron vessel or in a wooden tub, making first sulphide of calcium, by introducing steam through an iron pipe, to which a circular shape is given at the bottom with several little holes for the exit of the steam. After the fresh burned lime is charged, the water is admitted sufficiently to cover the lime several inches. The boiling commences immediately and the sulphur can be charged without delay. When the lime, during the stirring, seems to be in a soft, pulpy condition, it probably requires more water, which must be added, then sufficient steam can be admitted to keep the mass boiling, after which it will soon begin to assume a yellow color; water, if necessary, must be added to keep the lime sufficiently liquid. The boiling must continue from four to six hours. If the lime is of good quality, one and a half pounds of sulphur can be taken to three of lime. Polysulphide of calcium is formed, mixed with hyposulphite of lime and with bisulphide of calcium, and is now in the condition used to precipitate the silver. The boiling after four or six hours is stopped. After ten or twelve hours, the solution appears clear above the undissolved lime on the bottom.

It will soon be found out by practice how much water is required to obtain a solution of from 8 to 10 degrees Baume', and should not be less than 6° if intended for precipitation; but to convert it into hyposulphite of lime, 3° to 4° will answer.

The clear solution is then drawn off, either with a hose used as a syphon, or it can be discharged through several holes in the tub, one above the other, provided with short pipes. After all the clear liquid runs out, the tub is filled with water as high as it was before and boiled again. If this second solution should be too weak for precipitation, it can be used for dilution of another fresh charge of lime after the sulphur has been introduced. The remaining lime in the vat contains still some hyposulphite which may be obtained by filtration, but it filters very slowly.

The clear yellow liquid is now operated upon with sulphurous acid gas introduced by an iron pipe, and continued till the yellow color disappears, and a few drops of it do not make the slightest precipitation in a diluted solution of chloride of silver in hyposulphite of soda. It is now ready to be used as hyposulphite of lime.

The cheapest way to produce sulphurous acid under such pressure as to conduct it into a liquid through iron pipes, is to use sulphuric acid with charcoal. For this purpose an iron retort is charged with small charcoal (which would pass through a sieve of four holes to the running inch) to the half of its capacity; then as much sulphuric acid introduced as to form a pulp. After this the retort, surrounded by sand, is heated until the gas commences to escape through the pipe; the heat must be kept uniform. It is not necessary here to clean the gas.

From time to time sulphuric acid may be introduced into the retort, then again charcoal. The sulphuric acid is most suitable of 1.82 specific gravity.

Sulphide of Calcium,

the manufacture of which was described, together with the hyposulphite, gives a much better precipitation than the same salt of sodium. It precipitates quicker, and filters better; but the precipitated silver, if not washed with water before being pressed, retains the calcium, and is more difficult to melt. The sulphide can be kept in an open vessel for a long time, but the hyposulphite decomposes soon if not in use. Boiling the lime with sulphur in order to make sulphide of calcium, the quality of lime mnst be considered.

The sulphide is formed only from caustic lime, consequently more is obtained from fresh burned lime. Of this a certain quantity is charged into an iron kettle, water added, and then the pulverized sulphur. The proportion of sulphur and lime depends on the quality of the latter. The purest quality of lime from Santa Cruz, California, for instance, takes one pound of sulphur to 1.33 of lime. Of poorer qualities of lime it is better to take three pounds to one of sulphur and about ten parts of water. It is kept boiling for four or six hours, stirred with a wooden stirrer from time to time, and then allowed to cool and clear, and drawn off into wooden or iron tubs. In the residue, which may be drawn into a filter-box, there will be found more or less reddish-yellow crystals. These crystals are bisulphide of lime, and would serve likewise for precipitation, but they require about 400 parts of water to be dissolved, and in this condition the liquid is too weak to be of any use. The shorter the boiling time is, or if in proportion too little sulphur is taken, so much more of these crystals will be formed.

Working Auriferous Silver Ores.

Ottakar Hofmann's Patent Process. If rich auriferous silver ores, in which the percentage of gold is high, almost equal to that of silver, should be subjected to a chloridizing roasting, then impregnated with chlorine gas, leached with water, for the purpose of extracting the gold, and finally leached with hyposulphite of lime for the silver, it would in this case, although a high percentage of silver might be extracted, result in a yield of gold that would hardly amount to much more than about 50 per cent., more or less. The reason is not easily explained; the gold may be influenced somehow by the base metal chlorides during the roasting, which prevent the gold from being attacked by the chlorine gas.

On the other hand, if the base metal chlorides and the chloride of silver are extracted previous to the impregnation with chlorine, both metals—silver and gold—can be got out very close. On this fact is based O. Hofmann's gold and silver chlorination process, by which he successfully treated the gold and silver-bearing sulphurets of the Colorado No. 2, G. & S. M. Co., at Monitor, Alpine county, Cal.

The ore of Colorado No. 2, better known by the name "Tarshish mine," as it comes from the mine is poor, assaying in average from seven to nine dollars per ton, but the gangue, being decomposed feldspar, causes the ore to be soft and easily crushed, and is by means of Frue's concentrators very highly concentrated. The concentrated sulphurets consist of iron pyrites, silver glance, silver copper glance, fahlore, ruby silver, zincblende and galena. The value of the concentration varies a great deal, ranging from $250 to $1,700 per ton, with over 40 per cent. of the value in gold, which is not free in the ore. The average value of five months' run proved to be $564 per ton— $333.19 in silver and $230.84 in gold.

The operations of the process are as follows:

1. *The Roasting.* The concentrated sulphurets are subjected to a chloridizing roasting in a reverberatory furnace. These furnaces, although old-fashioned stirring furnaces, are considered by Hofmann preferable for this class of ores to any mechanical furnace in use.

especially to the continuous discharging ores. Those concentrated sulphurets require a very perfect roasting, and the furnaces, with continuous discharge, do not give time enough for very high sulphureted ore to become thoroughly desulphurized. Such high-grade ores require close attention, and the process must be under perfect control of the roaster. However, it does not take near as much time as required by Plattner's gold chlorination.

After the ore has been roasted, it is spread on the cooling floor and sifted, when cold, through a sieve of ten to fourteen meshes to the running inch. The sulphurets are heavy enough after roasting to make very little dust during the sifting, so that the inconvenient moistening can be obviated. The lumps are saved till a larger amount accumulates; they are then pulverized in a dry battery and slightly roasted.

2. *Leaching of Base Metals.* The roasted and sifted ore is charged into tanks with filter bottom, in quantities from two and a half to three tons, and leached with water to extract all soluble base chlorides. The water, if saturated with those base chlorides and some salt, of which generally some is left in the ore undecomposed after roasting, acts on the chloride of silver like a concentrated brine, dissolving the same. To prevent the escape of this dissolved part of silver, Hofmann does not admit the water from above the ore, as is usually done, but from below the filter-bottom, which, by means of a slight pressure, is forced to ascend through the ore to the top of the vat. In this way the concentrated solution accumulates above the ore, and in diluting the same by a stream of water and permitting the solution to flow out through the filter-bottom, the chloride of silver is precipitated on and through the ore, which is then extracted with the balance of the silver. This operation effects the fineness of the bullion somewhat, if there is a considerable amount of lead in the ore, but not materially. The bullion of Colorado No. 2 has by this process a fineness of 957.

3. *Leaching of the Silver.* This is done as usually by leaching the ore with a solution of hyposulphite of lime, and precipitating the silver with polysulphide of calcium. The hyposulphite dissolves more or less gold, so that the bullion of Colorado No. 2 contains from two to ten thousandths of gold.

4. *Second Leaching with Water.* After the silver has been extracted, the solution of the hyposulphite is allowed to run out till it disappears under the surface of the ore, when clear water is introduced again, in order to displace all solution. The desilvered ore is then removed from the tank to a dry kiln, where it is left for a time till the surplus water has evaporated. After this, it is charged back into the tank still moist. This second handling and drying cannot be avoided, as the ore after leaching is too wet to permit of a free passage of the chlorine gas, but if the works are arranged properly, this partial drying causes neither much delay nor much expense.

5. *Gold Extraction.* The gold in the ore is now in metallic condition, and very bright and clean, permitting a very close extraction. The rim of the tank is provided with a groove, which is open towards the inside, two and a half inches deep and one and a half inches wide. In this groove fits the cover of the tank, leaving, however, a play of one-eighth of an inch around the circumference. The cover being made of one-inch boards, the staves of the tank will project one and a half inches above the cover. This arrangement serves in two ways—first, it facilitates the operation to make the cover air-tight with clay, and second, it enables the chlorination to keep a sheet of water one inch thick on the top of the cover, thus making the cover perfectly air-tight and preventing the escape of gas into the working-room during the time of charging the tank with water for the purpose of extracting the gold. The cover is, furthermore, provided with two pieces of one and a fourth-inch gas pipe, six inches long, and a square opening six by six inches. During the time of the impregnation of

the ore with gas, these pipes are closed with balls of clay. As soon as the charge is ready for the extraction of the gold, these balls are removed and one of the pipes is connected with the hose of the water-tank, while the other, by means of a hose, either with another tank already prepared for chloridizing, or with the ash-pit of the roasting furnace. This is done to utilize the surplus of chlorine gas, and to protect the workmen from the very injurious effect of the same. Care must be taken to have sacks placed on the top of the ore right under the water pipe, and kept in place by two bricks to prevent the stream of water working into the ore. The square opening serves for exam-ining the progress of the gas in the ore, and can be closed air-tight by a good fitting cover and clay.

The chlorine gas is generated in a leaden gas generator, which is not heated, as usual, by direct application of fire, but by steam. For this purpose the generator is placed into a tight-fitting box, leaving a space of two inches around the side and bottom for the circulation of steam. The rim and cover are kept outside the box. The steam enters on one side through a half-inch pipe, while the other side of the box is provided with a one-inch exhaust pipe and an outlet for the water.

Wherever steam can be had in a chlorination work, this arrange-ment will prove very convenient and useful. The operator has the temperature entirely under his control. The least turn on the valve increases or decreases the heat, and, of course, regulates the genera-tion of gas. The discharge pipe of the generator projects a short way out of the box. On this pipe is fastened a piece of hose about two feet long, which can be closed with a thumbscrew clamp. The hose lies in a covered trough, which leads outside of the gas-house. In discharging the generator a small stream of water is permitted to flow through the funnel into the generator, by which the gas is forced through its usual outlet into one of the tanks. When the generator is filled, the stirrer is set in motion and the thumbscrew on the hose loosened. In this way the generator can be discharged without molesting or injuring the men. The cover of the generator is closed tightly with clay and only removed in case of repair.

The gas is conducted through a leaden pipe, intersected with rubber hose alongside and in front of the tanks, almost in the same height as the top of them. By means of a T and rubber hose, each tank is connected with the main pipe, and can be disconnected by the use of a thumb-screw clamp. The pipe through which the gas enters the tank is independent from that one through which the solution discharges. It is placed higher and as close to the false bottom as possible.

After the gas has been the proper time in contact with the ore, the gold is extracted by water, and precipitated in the usual way with sulphate of iron. The gold obtained is of extreme fineness, varying from 970 to 987 thousandths.

In treating very rich ores, containing say $700 to $800 gold per ton, the solution carrying out the gold is of a very lustrous yellow color, and the precipitated gold accumulates on the bottom is spongy lumps of great specific gravity, some of them showing scales of bright gold, which, under the microscope, might prove to be crystallized gold. There is but very little more time used in leaching rich gold ore than poor.

If the ore is copperous, considerable copper will be carried out with the gold solution, coloring the same green. In order to save the copper, the solution, after the gold has been precipitated and settled, is decanted into the copper tanks. But before doing this it is advisable to draw the solution first into a second gold tank, in order to catch the gold which should accidentally be carried off with the stream, and to leave it there for some time to let the gold settle again.

RESULTS.—To ascertain the working result of this process the concentrated raw sulphurets of the Colorado No. 2 obtained each day had carefully been weighed and assayed during a period of five months. The average value of these concentrations, as mentioned above, had been during said five months $564 per ton, with over 40 per cent in gold.

The total value of the bullion shipped at the end of this period, compared with the net value of the raw sulphurets worked during the same time, showed the actual working result to be, silver 96 per cent., gold 95 per cent.

Patera Process.

The most delicate operation in Patera's process is the preparation of the ore by roasting. The chloride of silver, formed during the roasting is dissolved by a cold solution of hyposulphite of soda, after all soluble base metals have been first leached out with hot water. Two parts of the hyposulphite of soda dissolve one part of chloride of silver, forming a soluble double salt. The tubs in which the ore is lixiviated with the hyposulphite of soda are small, receiving only 200 pounds of roasted ore. The extraction of silver is performed in the same way as described (page 104.)

Patera's process, to extract silver, cobalt and nickel, in the wet way, after an oxidizing roasting of six hours, is as follows: The roasted ore is placed in wooden vats with an addition of diluted sulphuric acid. Steam is introduced into the pulp and heated up to 100° F. The greatest part of the nickel and cobalt are now dissolved, while the silver remains unattacked. After the nickel and cobalt solution is drawn off or leached, diluted nitric acid is poured on the ore and again heated by steam, till the creation of red fumes ceases. This solution is also drawn off, the ore washed with hot water, and the silver precipitated by salt. Besides the silver there is also cobalt, nickel, some iron and arsenic, in this second solution. To make the precipitated chloride of silver fall quicker to the bottom, the liquid is brought into motion, and then after the solution appears clear, it is removed, by means of a syphon, and on account of some floating silver, permitted to stay quiet for twelve hours.

The chloride of silver, with diluted sulphuric acid, is conveyed into a vat, and precipitated in metallic condition by old clean iron. The silver is washed, dried, and melted into bars.

The solution containing nickel and cobalt is mixed with chloride of iron, and neutralized by addition of pulverized lime rock. Arsenate of iron and some iron oxide are thus precipitated; and the liquid, now purified, is condensed by evaporation. From this condensed liquid the cobalt is preciptated, by chlorate of lime, the solution separated, and the nickel precipitated by fresh burned lime. The nickel oxide is filtered, pressed, dried and calcined. After this it is ground fine, and then mixed with five per cent. of rye flour, and sufficient syrup to form a stiff dough, which is cut into cube pieces, dried, placed with charcoal powder into a crucible, and exposed to an intense white heat. The nickel oxide is reduced to metallic nickel ; the flour and syrup is burned off, and the nickel retains the cube shape, and appears brightly polished, after being kept revolving in a barrel with water for some hours.

Kiss Process.

This process extracts silver, and, to some extent, the gold. Roasting the ore, a part of the gold is transformed into such a state as to render it insoluble in water. After roasting, the ore is placed in filtering tubs and washed with water to remove the base metals. After this, a solution of hyposulphite of lime is conveyed on the ore, by which the gold and silver chlorides are dissolved and carried off into precipitating tubs. As soon as the sulphide of calcium is introduced, the gold and silver are precipitated as sulphides. The precipitation of both metals in a metallic condition is not admissible, for the reason that the hyposulphite of lime is decomposed if metallic copper is employed for precipitation. The results of Kiss's methods, practiced in Hungary, were not altogether satisfactory with concentrated sulphurets, but gave a good result with auriferous silver ores. It depends a great deal on the roasting whether more or less gold is extracted by the Kiss process. If a high heat is applied, then the greatest part of the gold is set free in metallic condition, and in this case very little gold can be dissolved by the hyposulphite of lime, because the metallic gold remains untouched, and only that part of it which during a slow cooling after the discharge, while the chlorine still escapes, happens to change into subchloride of gold, would be dissolved. If, however, the roasting is conducted at a dark red heat, just enough to chloridize the silver, then a larger part of the gold will combine as an oxide with soda and chloride of sodium, besides forming subchloride of gold, both combinations being insoluble in water, but dissolve in hyposulphite of lime.

Patera and Roeszner Processes.

The object of this process is, like that of the preceding, the extraction of silver and gold together. The ore is first subjected to a chloridizing roasting, by which the silver is converted into a chloride, while the gold remains mostly in metallic condition. The leaching liquid is prepared by conveying chlorine gas through a cold concentrated solution of salt to saturation. This chloridized solution dissolves silver, gold and copper at the same time. The roasted ore is charged into tubs with false bottoms, and the cold solution of salt and chlorine introduced. Silver ores treated after this method in Hungary gave 98.94 per cent. of the silver, all the copper, and nearly all the gold. An experiment on five tons of ore gave a clear profit of seventy-five florins, compared with the amalgamation.

Roeszner roasts the ore with salt, extracts a part of the silver by Augustin's method with a hot solution of salt, and treats the residue alternately with a solution of salt and chlorine, and with a hot concentrated salt solution for the extraction of gold and the remainder of the silver.

Augustin Process.

This process is not in use at present for silver ores, but for products of smelting. By this method the chloride of silver, which is formed by way of roasting, is dissolved in a hot solution of salt, and precipitated by metallic copper. One part of chloride of silver requires sixty-eight parts of salt to be dissolved.

Extraction of the Silver from Copper Matte and Black Copper. The principal aim with these materials is the oxidation of the copper as perfectly as possible, and then the chlorination of the silver. There are wooden leaching tubs of a small size—two feet eight inches in diameter, and nearly four feet high—fixed on wheels and arranged in one row. Into these tubs, which have false bottoms, the roasted stuff is introduced—about 800 pounds in each. Ores containing different kinds of earths cannot be lixiviated at a depth of over three feet; the metal oxides, however, allow the water to pass freely. This is also the case with roasted, concentrated, or pure sulphurets. Hot solution of salt is now allowed to flow through a trough in each tub. The salt penetrates the powder, dissolves the chloride of silver, and carries it through the filter at the bottom of the tubs, flows off to a reservoir, and from here, after the particles which may escape through the filter have settled, into a series of vessels, one above the other. These are provided with double bottoms. The two uppermost rows contain cement copper, six inches deep; the lowest, metallic iron.

The fluid deposits its silver principally in the first tub, dissolving at the same time an equivalent amount of copper. Some silver which escapes precipitation falls with the cupreous fluid into the next tub below, where the rest of the silver is taken up by the copper. In the third vessel the copper is precipitated by the iron. The brine, freed from silver and copper, is pumped up into the reservoir, heated and used again. The cement copper obtained in the last tub is placed back in the upper two. The brine circulates in the tubs until a bright copper plate is not coated with silver when held in the fluid from the leaching tubs. The residue, which is mostly copper-oxide, is removed, and an average sample taken and assayed. If it should contain over eight ounces per ton, it must be roasted over and again lixiviated.

The precipitated silver is taken out once a week and treated with muriatic acid, for the purpose of dissolving the copper particles which remained with the silver. After this, it is washed with water till all traces of the acid disappear, then pressed into balls, dried and melted.

Ziervogel Process.

Like the preceding, Ziervogel's extraction of silver is not applied to silver ores, but only to copper matte. The roasting is very delicate, and it is more difficult to obtain a satisfactory result with silver ores than by a chloridizing roasting. The silver in this process is converted into a sulphate, which is soluble in water, thus dispensing with the expensive salt brine. The pulverized and properly roasted copper matte is charged into leaching tubs, 500 pounds in each, and hot water admitted. As soon as the water begins to flow out, the hot water is made a little acid by admixture of some sulphuric acid. The lixiviation continues until a sample of the fluid remains clear if a solution of salt is added. The silver-holding brine is conveyed into a large reservoir, thirty feet long, where it clears of impurities, which accidentally come out of the leaching tubs, and falls from this reservoir through a series of cocks into the precipitating tubs. On the false bottom is a layer of cement copper, and upon this fifteen to twenty copper bars of 250 pounds weight. Each is fourteen inches long, five inches wide and one inch thick. The liquid loses most of its silver in these tubs, and flows then through a trough fifteen inches wide, lined with sheet lead and having a layer of copper pieces on the bottom, into five vats filled with copper, where the balance of the silver is deposited.

The desilverized brine comes now into a reservoir, whence it is pumped up into a large leaden pan and heated again by means of steam. Above this pan is a leaden vessel, out of which about thirty drops of somewhat diluted sulphuric acid drop into the liquid every minute. The acid prevents the separation of basic salts. The silver is taken out of the precipitating tubs every day. With it occur some copper and gypsum. The larger particles of copper are separated by washing, exposed for six or seven days to leaching with diluted sulphuric acid, and finally washed with hot water. The silver is from 860 to 870 fine. After drying, it is refined in a reverberatory furnace.

Once a year the brine is brought into contact with iron, in order to precipitate the copper. The purer part of the cement copper is used for the silver precipitation, and the finer part is delivered for smelting.

Extraction of Silver by Sulphuric Acid.

This method of extraction has not been applied to ores, but only to copper matte and copper alloyed with silver. The result is, precipitated metallic silver and sulphate of copper. It requires, however, a very perfect roasting, otherwise a part of the iron contained in the matte would make the sulphate of copper impure. It is, therefore, more proper to melt the matte for black copper and to treat this with sulphuric acid after granulation. The granulated copper can be dissolved in wooden tubs in diluted boiling sulphuric acid; the gold, if present in the copper, remains undissolved, and the silver is precipitated by copper plates. The liquid is then concentrated by evaporation to the crystallization point. A better economy in regard to sulphuric acid is obtained if the copper is first oxidized by being heated

9

in a reverberatory furnace ; but then it is generally the case that the copper must be repeatedly calcined, two or three times before all is dissolved. This inconvenience is avoided if, as before stated, the copper is subjected to boiling in diluted sulphuric acid.

Dissolving the copper in concentrated sulphuric acid is less advisaable ; it requires more sulphuric acid, as a part of it is used up for oxidation of the copper. The granulated copper is treated in iron or platina vessels with strong sulphuric acid of 64°. A great deal of sulphurous acid is created during the boiling, and this can be utilized by conveying it into leaden chambers for the purpose of converting it into sulphuric acid. From the solution the silver is precipitated by copper, but under some circumstances it may be more advantageous to precipitate the silver by sulphuretted hydrogen, which precipitates first the silver and then the copper. If the precipitation is stopped at the right time, sulphide of silver, with some copper, is obtained, which latter, after calcination, can be separated again by diluted sulphuric acid.

The Leaching Process.[*]

Under this name is understood a preparation of the ore applicable for the pan amalgamation. Its description, therefore, does not belong here strictly, but the leaching itself has so close a connection with the preceding manipulations, that this part alone may be described without mentioning the further treatment by amalgamation.

It is a well known fact that, in treating refractory ore in pans by amalgamation, of course by way of roasting, some very annoying things are encountered, and amongst them, principally, the great loss of quicksilver, amounting sometimes up to ten or twelve pounds per ton of ore; the rapid destruction of pans, which compelled many mills to use wooden sides fixed to the iron-pan bottom, a measure which saves the pans at the expense of quicksilver; and the very base bullion which results from such a treatment. In some instances it happens

---

[*]The Leaching Process is patented, as an application for pan and barrel amalgamation, by G. Kustel.

that a great deal, sometimes over fifty per cent., of iron goes into the amalgam, rendering the continuation of the amalgamation impossible. The result of the amalgamation of base metals is always a certain loss of silver which would have amalgamated if the base metals were out of the way. It happened very often in Nevada that $90 to $100-ore was purchased for the purpose of amalgamating it in pans; but a few tons proved that amalgamation had to be given up. Such ore is now considered suitable only for smelting.

At a very trifling expense all these difficulties can be avoided, and the amalgamation turned into a perfect success; for instance, the amalgamation of the silver ores at Flint, Idaho, turned out such base amalgam that further working proved to be ruinous. The introduction of the leaching process, however, resulted in a most favorable amalgamation. It is only to be regretted that after working several hundred tons, the mine refused to provide the mill with ore, perhaps on account of not having been sufficiently opened. The leaching for the pan amalgamation is most important, and at the same time cheap; all the expense is reduced to that of obtaining hot water. This process is not only important for silver ores containing base metals, but also for gold ores which by their nature require roasting. This refers principally to auriferous copper ores, as the amalgamation of gold is very much obstructed by the presence of copper salts.

It is a matter of surprise how simple a remedy could have been overlooked while fighting with the obstructions, caused by rebellious ores, during amalgamation. If there is soluble chloride of silver in the roasted ore, and besides this, soluble chlorides of copper, lead, antimony and zinc, it is a matter of course that all will be decomposed and amalgamated. All take part in consuming and parting the quicksilver, and in destroying the pan, hindering, at the same time, the easy amalgamation of the silver and gold. Why, then, not put all these obstructive metals out of the way, and give the silver a better chance to amalgamate? The base metal chlorides are soluble in water, the chloride of silver is not. It is therefore a most simple manipulation to dissolve those salts in water, and to remove them from the ore before amalgamation, by the leaching process. As soon as this is done the ore is divested of its rebellious nature, and it behaves in pans like the best ore. The process of leaching is described (page 104.)

IV. EXTRACTION OF GOLD

FROM SULPHURETS, ARSENIURETS OR QUARTZ BY CHLORINATION.

The extraction of gold without quicksilver is limited principally to that class of ores in which the gold is not free; that is, not in metallic condition, but combined with sulphur, arsenic or tellurium. Ores containing free gold, finely divided, the gangue of which is quartz without admixture of sulphurets, can be treated by chlorine gas directly without roasting, in the same way black sand may be treated, but sulphurets, arseniurets or tellurides have to be subjected to a thorough roasting. Silver ores, rich in gold, can be also advantageously treated with chlorine gas after a chloridizing roasting, provided that no coarse gold occurs in the ore. Ores or sulphurets containing coarse gold are not suitable for chlorination. By this process, if properly executed, 90 to 95 per cent. of the fire assay can be extracted.

In order to be sure of a result on a large scale, it is an easy matter to make an experiment with twenty or thirty pounds of sulphurets or of ore in the following way: The named quantity must be roasted first, and it is the most difficult task, requiring either a small furnace or a great deal of patience, especially when small charges are treated on a large piece of sheet iron, having a charcoal fire beneath. In either case the sulphur must be driven out perfectly, so that when in a glowing condition, no smell of sulphurous acid can be observed. When finished, a sample is taken for an assay, and the roasted stuff moistened with water, after the weight of the whole has been noted.

Fig. 17.

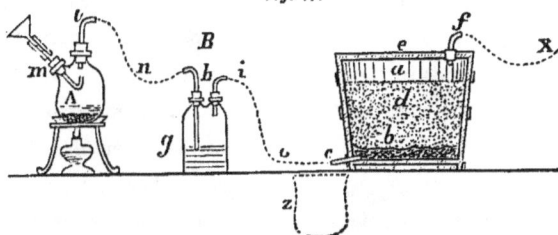

A common water bucket is then prepared to receive the moistened ore, which must not be too wet, but only moist enough to allow its being sifted. On the bottom of the bucket, *a*, Figure 17, some clean rock or broken glass is placed about two inches deep, and covered with a piece of moistened canvas. A short glass pipe, *c*, a quarter of an inch in diameter, is inserted close above the bottom.

The ore, *d*, is then introduced, filling up two-thirds or less of the space as loosely as possible, and covered with a wooden or iron cover and pasted all around with dough. The cover is provided with a short glass tube, like *c*, to which an india rubber tube, *f*, for carrying the gas out of the room, is attached. Both glass tubes, *c* and *f*, must be likewise secured with dough.

The chlorine gas is generated in a glass vessel, *A**. There are two corks in it, each having a glass tube, as represented in the drawing. The cork, *l*, is removed and the vessel charged with 3 ounces of pero-

---

* All materials necessary for such an apparatus can be procured from John Taylor & Co., 118 and 120 Market street, and 15 and 17 California street.

side of manganese, 4 ounces of common salt, and $4\frac{1}{2}$ ounces of water, all well mixed. The cork is inserted again and well secured with dough. Another vessel, *B*, provided with two necks, contains water, as indicated by *g ;* the glass tube, *h*, dips about one-half inch into the water. The corks are made air-tight like the others in *A*. The whole apparatus is now joined together by rubber pipe, *n* and *o*, fitting tightly to the glass tubes. Having all thus prepared, $7\frac{1}{2}$ ounces of sulphuric acid are poured through the safety-tube, *m*, but only in small portions and at intervals. When the bubbling of the water at *g*, in the vessel *B*, is not lively enough, some more acid is introduced, and finally the temperature raised by an alcohol lamp. If all the joints have been luted carefully with dough, not the slightest inconvenience will be met with. The chlorine gas from the generator, *A*, is forced through the water in *B*, by this means washed from muriatic acid. Through the pipe, *o*, it enters the bucket and ascends slowly till it reaches the cover, escaping then through the rubber pipe, *F*, where it must be examined from time to time by dipping a glass rod into ammonia and holding it to the end of the pipe, *x*, which leads out of the room. In contact with chlorine the ammonia evolves white fumes, and chlorine can be detected by these means wherever it may escape. The gas is allowed to pass through the bucket as long as chlorine is created. In this condition, by stopping up the pipe, *x*, if no more chlorine is evolved, the apparatus may stand until the next day. The cover is then removed, the pipe, *o*, taken off, a clean glass or porcelain vessel, as indicated by *z*, placed below *c*, and cold water carefully poured over the ore till the bucket appears to be full. The solution which comes out at *c*, must be examined at times in a small tumbler with a few drops of a solution of sulphate of iron. If the clear solution remains unchanged, without becoming darker, the lixiviation is finished.

To the solution in the vessel, *z*, a few drops of muriatic acid, and then sulphate of iron, or green vitriol (dissolved), is added and stirred with a glass rod. The whole is allowed to stand till all the gold is precipitated and the liquid is perfectly clear. This is drawn off by means of a syphon, for which the rubber pipe, x, can be used. The remaining fluid and the precipitated gold are gathered on a filter, washed with warm water and dried with the filter in a porcelain cup, above an alcohol lamp. The filter is burned either free or under a muffle, care being taken not to lose a particle of the filter ashes; mixed with some lead it is then cupelled and the gold button weighed. A comparison with the assay shows to what percentage the chlorination has proceeded.

The same experiment can be made also with from five to seven ounces of sulphurets in a glass cylinder, or other shaped bottle, as Figure 18 shows, the bottom of which is cut off by means of a glowing piece of charcoal, or otherwise.

FIG. 18.

The roasting may be performed over charcoal or coke in a small stove on a piece of sheet-iron, the edges of which are bent up and the inside coated over several times with clay-water, and then well dried, or the roasting can be done in a larger black lead crucible, through the bottom of which a hole is cut for the draught. The sulphurets, finely pulverized, must be stirred with an iron spatula until no smell of sulphurous acid is perceptible, after which a strong red heat is applied. When cold, the sample must be ground over in an iron mortar and roasted once more at a red heat. When cold, it is moistened in a dish or cup with sufficient water, to make it of a loose or wooly consistency.

If the roasting is perfect, the metallic gold could be dissolved by leaching with chlorinated water, but if some sulphurets remain undecomposed, it is more proper to use chlorine gas. The pint bottle, or a similar glass vessel, is prepared with a cork in the neck. Through the cork is fixed a glass tube, to which a rubber pipe is attached through which the gas is admitted and which serves also for the discharge of the solution. The neck is filled with fragments of clean quartz or broken glass covered with a piece of cloth. On this the moistened ore is placed and proceeded as described in using the bucket.

Treatment of Sulphurets by Chlorination.

This process, invented by Plattner, was first successfully introduced in Silesia, Germany, in the year 1851, on auriferous arsenical residue, which had accumulated for many years. With various modifications the process was practiced afterwards in Saxony, Hungary, Transylvania, etc., and, in 1858, in California. This process is very perfect if well executed. The extraction of gold from roasted sulphurets by amalgamation in pans is very difficult, resulting always in a great loss, while, if treated by chlorine, the same kind of pyrites will yield 90 per cent. of the gold contained, or more.

The operations to which concentrated sulphurets or arseniurets are subjected are principally the following : First, the oxidizing or chlori-dizing roasting as explained, (pages 21–52). Next, impregnation of the roasted ore with chlorine gas in wooden vats, then leaching of the soluble gold with cold water, and finally precipitation of the chloride of gold by sulphate of iron. The finer the gold particles are the easier and quicker the chlorination will be finished. The presence of copper, zinc or antimony does not interfere with the chlorination ; lead ore, if too much, may prevent a perfect roasting, and silver is only then injurious if alloyed with gold in too large a proportion, unless it is very fine. Generally, the silver is not extracted, for the reason that the amount is too insignificant to admit of a remunerative extraction ; but if, besides the gold, so much silver should occur in sulphurets as to make it an object of saving, it can be leached out with hyposulphite of lime, after the gold has been removed, and it will be found that this silver will often contain so much gold as to increase the value of an ounce by 20 to 25 cents, although no more gold could be leached out with water. The better result, obtained by O. Hofmann's process, extracting the silver before the gold is chloridized, is described, page 122.

Coarse gold requires too much time to be converted into a chloride.

The presence of lime and talk (silicate of magnesia) makes the chlorination of roasted ore troublesome, but an addition of one to three per cent. of salt, according to the amount of those earths, removes the difficulty; but for all that, it sometimes happens that, in precipitating the gold, other substances are also precipitated, which will be mentioned further on.

The roasting is a very important part of the whole operation; on this the success of the chlorination principally depends. Concentrated sulphurets, if there is no opportunity of being worked directly, should be kept under water, for, if exposed to a slow drying, and to getting wet alternatively for several months, it forms hard lumps and crusts, and must be pulverized anew before charged into the furnace.

The construction of the furnace has but little influence on the result of roasting. A perfect desulphurization is the first condition. This requires time; and using a reverberatory furnace, the time can be given *ad libitum*, 10 or 30 hours, as the case may require; the same refers to the cylinder furnace, roasting by charges. Furnaces like a retort, or a muffle furnace, in which the sulphurets are exposed to the oxygen of the air without coming in contact with the flame, give no better results, and the roasting is more expensive on account of the consumption of more fuel.

In regard to the loss of the gold, it has been stated that, in some places the loss amounted to a very high percentage, even to nearly one hundred per cent.; this, however, was not confirmed by the experiments of Plattner, who exposed diverse combinations of gold

with sulphur and arsenic, different ores with and without free gold
in the finest state, to different degrees of heat and time of roasting,
and the result showed that a loss of gold occured only when the roasting was performed so rapidly that the finest gold particles were carried
off by the volatile products of roasting. Many other experiments
showed that the loss of gold is inconsiderable. The addition of salt
to the sulphurets or arseniurets has no influence on the loss of gold
either.

Application of Chlorine Gas.

After the ore has been properly roasted (page 56), it must be moistened before introduced into the chlorinating vat. The sulphurets, if
dry, form a more condensed mass than in a certain damp condition,
and are, therefore, more obstructive to the ascending gas; besides, the
chlorine does not attack the dry ore as vigorously as when it is damp.
It is, therefore, indispensable to moisten the sulphurets after they are
sufficiently cool. For this purpose the roasted charge, or several tons,
are spread on the floor and water sprinkled over it by means of a
hose, or otherwise, and turned over several times till it appears uniformly wetted. The moistened charge should not create the slightest
dust, and at the same time the hand should remain dry on handling
it. A handful of it pressed hard together must form a lump, which
can be held with the fingers, but falls into its former loose condition
if handled. This moistened ore always contains some lumps formed
in the furnace; sometimes it may happen that a nail or screw is
charged with the ore. Iron is very injurious, if in the ore during the
chlorination, and the lumps would not chloridize properly. For this
reason, sifting of the moistened ore is necessary. The separation of
lumps and other impurities is not the only reason why sifting should
be done. The more chlorine gas that can be introduced into the ore-
box or vat, the surer will the chlorination of the gold be effected; it
is, therefore, important to sift the ore directly into the vat in order

to have it therein in as loose a condition as possible, leaving more space between the ore particles for the chlorine gas. The sieve, about eight meshes to the running inch, is twelve to fourteen inches by twenty-five in the clear, the sides five inches high.

The chlorination vat, into which the ore is sifted, by pushing the sieve to and fro, either on two scantlings laid over the rim of the vat or suspended on four ropes, is seven to ten feet in diameter, capable of holding four to six tons. These vats, or square boxes, are provided with false bottoms, in order to permit leaching and to allow the gas to expand below the whole mass of ore, and to ascend through the ore uniformly at all points. The arrangement of the filter bottom can be fixed in the same way in a vat, as shown in a square box, Figure 14, page 105; if preferred a cotton cloth for a filter, but it is more expensive than a gravel filter, the cloth being soon destroyed by the chlorine. In case cloth is used, it is necessary to lay staves across over the cloth four inches apart, so that when the tailings are taken out by means of shovels, the cloth should not be reached and dis placed.

Fig. 19.

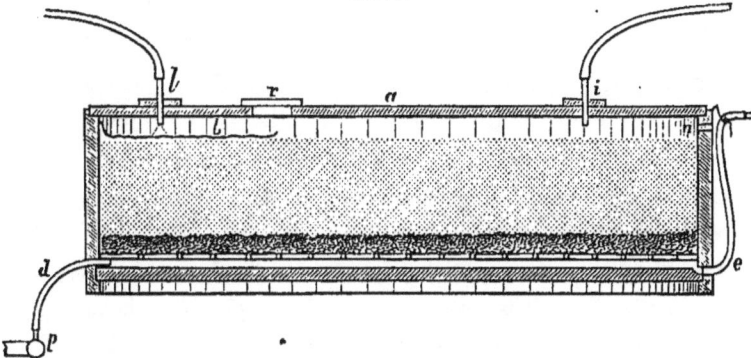

Figure 19 represents a tub filled with ore on a gravel bottom. Over the filter bottom, which is perforated with inch holes (or direct on the bottom of the tub), is spread, first a layer of clean quartz, one and a half to two inches square. In default of quartz, another kind of rock or gravel will answer the purpose, except lime rock, which absorbs a great deal of chlorine. Over the coarse layer smaller pieces are laid, and so on, decreasing in size till a layer of sand covers the whole, forming thus a filter of from four to five inches in thickness. This filter remains always in the vat; the shoveling out of the residue, therefore, must be done carefully on approaching the filter bottom. There are two holes communicating with the space below the filter bottom. One is for the reception of the lead pipe, *d*, or better, a rubber hose, by which the chlorine is introduced; the other, *c*, is provided with a hose for the discharge of the lixiviation. This side of the tub stands half an inch lower than at *d*.

To prevent the absorbtion of the gold solution by the wood, it is necessary either to keep the vat filled with salt brine for several days, to fill up the pores of the wood, or to coat the inside of the vat with asphaltum varnish, or a mixture of pitch and tar or oil. The pitch mixture must be applied while hot; the asphaltum varnish is preferable, as it penetrates the pores more perfectly than the pitch. In either case, however, after a long use, when the vat should be discarded, the staves should be burned and the ashes assayed for gold.

The moistened ore is then sifted into the vat, within about six inches of the rim, as seen in the drawing, leveled carefully, without pressing. The wooden cover, *a*, is laid upon the rim, after a piece of cloth, *b*, has been tacked to the side of the vat near the water-pipe, *l*, to prevent the jet of water disturbing the surface of the ore. The pipes, *i* and *l*, before the hose is used, must be closed with stoppers, also the hose, *e*. It is better, in place of the pipe, *l*, to have only a hole into which the hose is inserted, when the water is needed. The chlorine gas can be admitted now into the impregnating vat.

At Melrose, six square boxes, six by five feet, prepared on journals, admitted a very convenient and quick discharge.

Fig. 20.

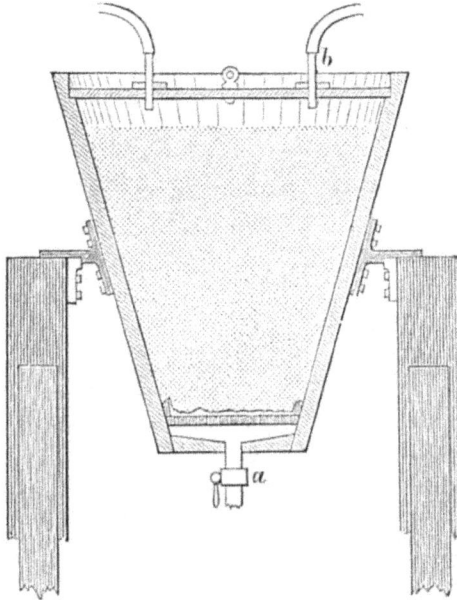

Fig. 20 shows a vat that was planned by E. Riotte, for chlorination works in Japan. Vats of a similar shape were also used in Germany to chloridize the sand in the river Rhine. It is a deep vat, but since the roasted sulphurets filter always very readily, a column of sulphurets four or five feet deep still permits a good leaching. The vat measures four feet on top and two feet at the filter bottom, and is five feet deep. This arrangement has the great advantage of taking up less room, and having comparatively a much smaller fitlteing surface, the more convenient cotton cloth can be laid over the false bottom, which is perforated. But also a gravel filter between the bottom and perforated partition can be prepared, as described in Fig. 19, which does not get out of order by being turned over. The gas is admitted through the lead pipe, a, provided with a cock, by which the pipe is closed, after the required amount of gas is passed into the

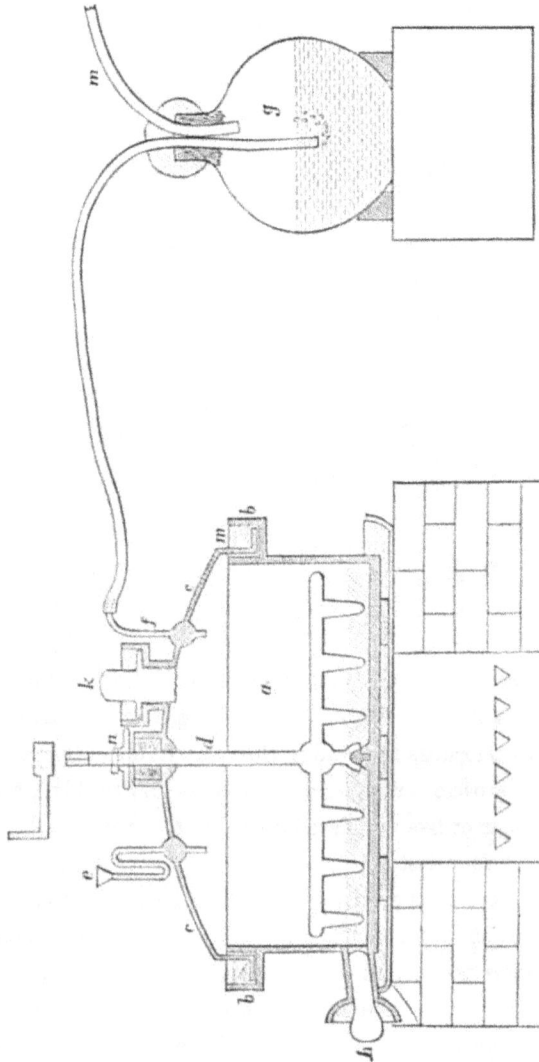

Fig. 21.

tub. The same pipe, with another hose attached, serves for the dis-charge of the lixivium. The cover is represented in the drawing, after O. Hofmann's design. This arrangement makes the cover perfectly air tight. After the joint on the periphery has been carefully luted with a good plastic clay, also around the two pipes, the whole cover is then covered with water about one inch deep. The clay, under water, of course cannot crack, and this is an advantage, especially then, if before leaching it were found at one of the pipes, that no surplus chlorine gas is in the tub and a second charge of gas is required.

This gas (page 14) is produced in a leaden vessel, as shown in Fig. 21. The circular tub, a, has on the outside a circular ring, b, two and a half inches deep and two inches wide, for the reception of the flange of the cover, c. The cover is provided with a center hole for the reception of the stirrer, d, with a leaden pipe, e, bent in the shape of the letter ∞ through which the sulphuric acid is introduced ; another pipe, f, conveys the gas to the washing apparatus, g. At the bottom of the gas generator is an opening, h, for the discharge of the residue, and k serves for introducing the charge.

The cover is placed over the generator, the edge of the flange luted with clay, as shown by m, all round, and then filled with water, as indicated by the dotted line. The space at n, round the stem of the stirrer, must be filled with packing of cotton cloth strips saturated with oil,* and the outlet, h, closed with a wooden plug wrapped like the stuff-box in cotton cloth, also saturated with oil, and the space between collar and plug luted with clay, or better, with putty. The stirrer rests on a glass button, inserted in the cavity of the lead socket.

All roasted sulphurets do not consume the same amount of chlorine gas. The following approximative charge may serve to chloridize three tons :

---

* It is for the manipulation with chlorine gas, a very important discovery, made by E. Riotte, that oil, in contact with chlorine, forms a sticky substance, which prevents the chlorine from penetrating it ; and the india rubber hose, if oiled inside before use, retains its flexibility, especially if the oiling is repeated from time to time. It is thus very easy to stop the escape of chlorine gas wherever leakage is observed, by applying some suitable material, impregnated with oil. It would also give an excellent coating, a substitute for pitch or asphaltum, if the chlorination vats were coated two or three times with warm oil, which enters readily the pores of the lumber much deeper than asphaltum varnish.

          30 ℔s. of manganese (peroxide) pulverized,

30 to 40   " " common salt, according to quality,

     75   " " sulphuric acid of 66° B., and

     45   " " water ; or,

          30 ℔s. of manganese,

     30   " " salt, good quality,

     85   " " sulphuric acid 63° B., and

     35   " " water.

Sulphuric acid of 66° costs 4 cents per pound in carboys, and the same of 63° $2\frac{1}{2}$ cents ; the latter charge, therefore, is cheaper, unless the freight would exceed seven cents per pound.

Water, salt and manganese are first introduced, and the charge hole, k, closed with a wrapped solid plug of lead, as above described, and luted with clay. The generator is placed in a square piece of boiler-iron, the edges of which are bent up two or three inches to hold water or prepared for steam. The sulphuric acid is now introduced through the pipe, e, but not the whole amount at once ; about twelve pounds to begin with will answer to create sufficient head to develop the gas, which escapes through f into the carboy g, to be washed from acids. The gas should bubble through the water very lively; if not, the stirrer should be turned, and if this fails to produce the required effect, more sulphuric acid must be introduced and repeated as often as the bubbling appears to become less active, always using the stirrer first. After all the sulphuric acid has been used up, the fire must be started till the water below the generator commences to boil. It takes about four hours to develop all the chlorine gas.

The rubber hose from the pipe, f, ($\frac{3}{4}$ inch hose), after being oiled inside, is inserted into the carboy one or two inches below the water ; the other hose, m, conveys the gas to the chlorinating vat. Both rubber pipes in the neck of the carboy are wrapped in oiled cotton cloth, and all the space around packed tight with oiled rags, then covered with putty, as shown in the drawing. There is no necessity to change the water in the carboy often—once every month will do. The water through which the gas passes absorbs, if cold, about two and a half times its volume of gas, and is then saturated, but is still good for taking up muriatic acid.

The washing of the gas, to get rid of the small quantity of muriatic acid, is not so important, since, if a portion of this should happen to enter the ore, and forming sulphureted hydrogen, precipitate metallic gold of that already chloridized, this would again be converted into a chloride in presence of abundant chlorine; but the washing is an indispensable indicator of the progress in the generator, by the more or less lively bubbling. Care must be taken to turn the stirrer from time to time, while the water round the generator is boiling, to prevent the caking of the ingredients.

If there is only one chlorinating vat, the gas from the carboy can be conveyed through one hose direct into the vat below the filtering bottom; but if there are several of these, it is better to fix a lead ($\frac{3}{4}$-inch) along the vats, connected with those half-inch pipes before each vat, to which a piece of hose is attached, as seen in Fig. 19. If not in use, the hose is withdrawn from under the filter, and closed with a stopper.

After the hose, d, has been inserted under the false bottom, the gas enters the space and rises slowly through the ore, until it reaches the cover, and commences to escape through the pipe, i, or in Fig. 20, through b. This is easily detected by the smell, but it frequently happens that although the smell denotes the escape of gas, it is sometimes difficult to find the point where it escapes. Ammonia, in contact with the smallest quantity of chlorine, forms a white fume, and it is the best medium to examine all connections therewith to find the escape of gas. If, in the generator,* little gas is still emitted, this

---

* At Melrose, a gas generator of a different arrangement was successfully in use. In shape and construction, it was exactly made like the carbonic acid generators for the fabrication of soda water. The cylinder is cast lead, and has for protection several rings of boiler iron on the outside ; only pure lead can be used as solder,

may continue; also, if all outlets in the chlorinating vat are closed. After this, water is admitted into the generator through the pipe, f, Fig. 21, to fill up the space inside, and to displace the gas. The plug, h, is then removed and the residue discharged into a closed trough.

The hose, d, in figure 19, is also removed and both closed; the hose, c, and the hole at d, the latter well luted with putty or clay. In figure 20, the connection is cut off by the cock, a. In case the roasting is not properly done, it takes several pounds more of manganese and the other materials in proportion, as well as more time for impregnation.

Lixiviation.

After twelve hours, or if the sulphurets contain coarse gold, after fifteen or eighteen hours, the little cover, r, is taken off to see whether there is enough of surplus gas in the vat; if only very little is observed, the result of chlorination is not certain, but if all gas was consumed, it is best to start the generator and impregnate anew. Holding before the opening, r, a few drops of ammonia on a rag, it must show a thick white fume. The opening is closed again, and the hose, i, is led either out of the building, or in another vat already prepared for impregnation. The stopper of the hose, c, is removed, and then water let in through l. The upper end of the hose, e, is inserted through a hole, n, below the cover into the box, to let the gas escape through the pipe, i. The water should flow in quickly, until it nearly reaches the hole, n. It is a great advantage to use at least one inch hose for the water. The hose, e, is now lowered into a gutter leading to the precipitating vat, which may be made like the vat for precipitating the silver (page 114, figure 15); but not higher than three or four feet, and well coated with asphaltum varnish, or it may be made as a rectangular box three feet deep, and lined with sheet-lead, half round on the bottom, which permits an easier and better cleansing. A very smooth surface in the precipitating vat is important, to facilitate gathering the finely precipitated gold. There should always be three to four of those precipitating vats, because the water sometimes shows signs of gold after one or two of the precipitating vats are already filled.

As the filtering water flows from the gutter into the precipitating tub, samples should be taken frequently in a clean tumbler of white glass, and observed whether a few drops of sulphate of iron (green vitriol) cause a dark, or bluish precipitation. If this be the case the leaching must continue till a sample under such examination remains perfectly clear. After this, the water supply in the chlorinating vat must be stopped and all the liquid contents of the vat permitted to flow into the precipitating vat.

Precipitation.

The precipitant for gold is a solution of sulphate of iron. It is known also by the name of copperas, or green vitriol, and forms light green crystals. Dissolved in water it generally makes a muddy solution, but it clears in a few hours, while a sediment falls to the bottom. It is, however, better economy to prepare the sulphate in the chlorination works, by placing old iron into a tub, with from five to six buckets of water, and fifteen or twenty pounds of sulphuric acid in two or three charges at intervals of two or three hours. There must be always a surplus of old iron in the tub. This is prepared two or three days before the solution is wanted. One bucket of this solution, or less, according to the richness of the ore, is poured into the precipitating tub, and the liquid stirred well. The precipitated gold shows a red brown color, but there may be a great deal of gold yet in the liquid, not precipitated. Whether more of the sulphate is required can be ascertained by taking a sample of the liquid, filtering it through filtering paper, and mixing it with several drops of the sulphate of iron solution. If the mixture should darken a little, after a few minutes, it would show that more of the precipitant is needed. The precipitated gold requires some time, from 12 to 15 hours, before it is deposited on the bottom. The fluid above the gold must appear perfectly clear before it can be drawn off.

The water which is drawn off, although apparently perfectly clear, still it may have suspended some gold ; it is therefore very important to convey it into a large tub, where it is allowed to remain till the next charge is drawn. From this vat the water must flow through two other vats with filtering bottoms, which are covered with three or four layers of cotton cloth. Also sawdust, or from six to eight inches tailings from roasted sulphuret tailings, are used as filters for this purpose. After several months of operation these filters may be examined, and if rich enough, the gold extracted by chlorination with gas or leached with chlorine water and precipitated with iron solution.

The precipitated gold in the precipitating vat is not taken out before it is accumulated from a longer run; it is more profitable to make as long a run as possible, as there is a smaller percentage of loss by waste in handling gold in large quantities. This is dipped out carefully by means of a dipper or scoop into a clean porcelain dish or enameled vessel, and then the tub washed clean with a brush and a jet of water on the sides and bottom. The gold and washing is then introduced into a filter of filtering paper, or of cotton cloth, washed well with hot water, dried, and finally fused in a black lead crucible, with addition of borax and some saltpetre.

If the sulphurets are pure iron sulphurets, or iron pyrites, with some quartz, the precipitated gold is of a fine red brown color, filters and fuses easily without any trouble, but it is different with impure sulphurets. Of these the leaching after chlorination appears clear and of the proper yellow color, but on addition of sulphate of iron the precipitation often appears of a blackish, bluish, or whitish color ; the precipitated gold is voluminous, it filters very slowly, and in smelting a great deal of borax must be added. The impurities are mostly iron, sulphate of lime, etc. A great deal of these impurities can be dissolved by pouring from five to ten pounds of sulphuric acid into the precipitating tub after the clear water has been removed, but it is much better to boil the gold in an enameled kettle with the acid, then dilute it and filter. The gold will be still mixed with foreign matter. In many similar cases a better result is obtained if, immediately after leaching, sulphuric acid is added, well stirred, and the next day the clear liquid drawn over into another precipitating vat and the gold precipitated with sulphate of iron.

In Europe sulphureted hydrogen is used to precipitate the gold as a sulphide; the sulphureted hydrogen is obtained there by heating matte from lead, or smelting, with diluted sulphuric acid. A convenient and cheap way to produce that gas is to heat in a proper glass vessel paraffine and sulphur flour. The precipitation with sulphate of iron is preferable.

The cost of chlorination and the roasting in a long furnace of two tons of sulphurets per twenty-four hours, calculated on San Francisco prices of materials, the sulphuric acid in carboys, is as follows :

Superintendence	$ 6 00
Two roasters, @ $2.50	5 00
Two cords of wood, @ $5	10 00
20 lbs. of peroxide of manganese, @ $2.75	0 55
27 lbs. of salts (ground), @ 50 cts	0 13½
57 lbs. of sulphuric acid 63°, @ $2.50	1 42½
One man, at the rate of $2.50	2 50
Sulphate of iron, about	0 29
Total	$25 90

Or $13.00 per ton of sulphurets.

The presence of galena necessitates a good roasting, and a strong finishing heat, as far as the fusibility of the metal permits it. Ore of this kind requires a long furnace in order to prepare the ore for the stronger heat while it progresses from one hearth to the other. In case the ore is not properly finished, the undecomposed sulphurets and sulphates would absorb a great deal of chlorine. An experiment made with two samples of ore, the one contained three per cent. of galena, the other thirty. After roasting without salt, both were subjected to impregnation with chlorine gas. The first ore consumed the usual quantity of gas, and gave 85 per cent. of the gold contained, but the tailings showed coarse gold particles which could not be chloridized in twelve hours. The other lot absorbed twice as much chlorine, was leached after 24 hours, and the result was 62 per cent. of the gold. The tailings contained, after grinding, specks of undecomposed galena, which proved that the roasting was either too short or there was not sufficient heat. The coarse gold was taken up instantly by quicksilver in consequence of the clean surface created by the action of chlorine, but if in such a case quicksilver should be used, the tailings ought to be washed or the leaching continued till all chlorine disappears, and the tailings amalgamated while yet fresh, because if too long exposed to the action of the air, the gold particles soon appear reddish and do not readily amalgamate.

The question in regard to utilization of the surplus gas which remains in the vat after the chlorination has been finished, seems not to be solved yet by practical use. A vat of seven feet diameter and two feet high, when filled with roasted ore to within six inches of the top, leaves about forty-seven cubic feet of space for the gas. This space should be always filled with chlorine gas, and should be replaced if, before leaching, it were discovered that all the gas was consumed on account of improper roasting, or from some other cause. Having then two impregnation vats, the communication between them, for the purpose of conveying the gas from one into the other, is easily prepared by a hose from the top of the one connecting under the filter bottom with the other vat. The admitted water will force the gas into the other vat, but as the cold water dissolves a great deal of the gas, perhaps more than the half of the whole amount, which for the extraction of the gold is advantageous, it. is hardly worth while to utilize the remainder of the gas, if it were not for the object of getting rid of it in an inoffensive manner.

At Schemnitz, Hungary, the tailings from the silver extraction by Ziervogel's method, are treated for gold by the chlorination process. They contain about three ounces and a half of gold per ton. The gas generator is made of cast iron with a leaden hood, joined by flanges and bolts. The hood has three openings, for charging manganese, for sulphuric acid, and the third for the exit of the gas. This enters first the washing vessel, then a receiver, and thence passes through two pipes into earthen impregnation vessels. The lower part is contracted like a funnel, and receives a filter of quartz some two inches deep. The upper part is furnished with a clay cover provided with a suction pipe.

The tailings, having been dried and slightly glowed, for the purpose of destroying the basic salts, are moistened in the usual way, and six hundred pounds introduced into each of the four chlorination vessels. The chlorine is then introduced through a pipe below the quartz filter and continued until the gas is distinctly perceived above the charge. The cover is then put on, and luted with dough. ¡The tailings are exposed to the action of the chlorine for twelve hours. At the end of this time warm water is introduced, and the lixivium collected in large glass bottles, where the gold is precipitated by sulphate of iron.

Colvert's Method for Auriferous Quartz.—This method is based on the production of chlorine in the mass of ore. It is cheap, and permits of the extraction at the same time of both silver and copper, and is not injurious to the workmen. The finely pulverized quartz is mixed with one per cent. of manganese, charged into vats and muriatic acid added. In this condition the ore is kept closed in the tubs for twelve hours. There is a perforated filter bottom in the tub covered with some brush and straw. The mass is then leached several times with the same water, and then conveyed into precipitating tubs, where the copper is first obtained by means of iron, and afterward the gold, the chlorine having been driven off by heating the liquid. The precipitant is sulphate of iron.

If there is silver in the ore, the chlorine is generated with salt, manganese and sulphuric acid taking six parts of salt to three parts of manganese. The formed chloride of silver dissolves in the salt solution and is precipitated by copper plates, the copper by iron, and the gold by sulphate of iron.

Other Publications

by

Miningbooks.com

- **Placer Gold Deposits of Nevada**
- **Placer Gold Deposits of Utah**
- **Placer Gold Deposits of Arizona**
- **Gold Placers of California**
- **Browns Assaying**
- **Arizona Gold Placers and Placering**
- **Arizona Lode Gold Mines and Gold Mining**
- **Dredging for Gold in California**
- **Metallurgy**
- **Gold Deposits of Georgia**
- **Placer Examination: Principles and Practice**
- **Geology and Ore Deposits of the Creede District, Colorado**
- **Gold in Washington**
- **Placer Mining in Nevada**
- **Gold Placers and their Geologic Environment in Northwestern Park County, CO**
- **Placer Mining for Gold in California**
- **Geology and Ore Deposits of Shoshone County, Idaho**
- **Gold Districts of California**
- **Gold and Silver in Oregon**
- **The Porcupine Gold Placer District Alaska**
- **Gold Placer Deposits of the Pioneer District Montana**
- **Economic Geology of the Silverton Quadrangle, Colorado**
- **The Ore Deposits of New Mexico**

- Roasting of Gold and Silver Ores, and the Extraction of their Respective Metals without Quicksilver

- Geology and Ore Deposits of the Summitville District San Juan Mountains Colorado

DEWEY & CO'S